VOLUME ONE HUNDRED AND SIX

ADVANCES IN
COMPUTERS

VOLUME ONE HUNDRED AND SIX

ADVANCES IN
COMPUTERS

Edited by

ALI R. HURSON
Missouri University of Science and Technology,
Rolla, MO, United States

VELJKO MILUTINOVIĆ
University of Belgrade,
Belgrade, Serbia

ACADEMIC PRESS
An imprint of Elsevier

Academic Press is an imprint of Elsevier
50 Hampshire Street, 5th Floor, Cambridge, MA 02139, United States
525 B Street, Suite 1800, San Diego, CA 92101-4495, United States
The Boulevard, Langford Lane, Kidlington, Oxford OX5 1GB, United Kingdom
125 London Wall, London, EC2Y 5AS, United Kingdom

First edition 2017

Notices
Knowledge and best practice in this field are constantly changing. As new research and
experience broaden our understanding, changes in research methods, professional practices,
or medical treatment may become necessary.

Practitioners and researchers must always rely on their own experience and knowledge in
evaluating and using any information, methods, compounds, or experiments described
herein. In using such information or methods they should be mindful of their own safety and
the safety of others, including parties for whom they have a professional responsibility.

To the fullest extent of the law, neither the Publisher nor the authors, contributors, or editors,
assume any liability for any injury and/or damage to persons or property as a matter of
products liability, negligence or otherwise, or from any use or operation of any methods,
products, instructions, or ideas contained in the material herein.

ISBN: 978-0-12-812230-3
ISSN: 0065-2458

For information on all Academic Press publications
visit our website at https://www.elsevier.com/books-and-journals

Working together
to grow libraries in
developing countries

www.elsevier.com • www.bookaid.org

Publisher: Zoe Kruze
Acquisition Editor: Zoe Kruze
Editorial Project Manager: Shellie Bryant
Production Project Manager: Surya Narayanan Jayachandran
Senior Cover Designer: Christian J. Bilbow

Typeset by SPi Global, India

CONTENTS

PREFACE

Traditionally, *Advances in Computers*, the oldest series to chronicle the rapid evolution of computing, annually publishes several volumes, each one typically comprised of four to eight chapters, describing new developments in the theory and applications of computing. The 106th volume entitled "Education in Computing and DataFlow Super Computing" is a thematic volume inspired by the advances in computer architecture in general and more specifically in DataFlow processing. In addition, the volume includes a chapter summarizing a collective experiences of the faculty of several departments of the University of Belgrade, as far as the development of a graduate course that teaches the students about all the skills necessary to bring a sophisticated research and development project to its end, in general and specifically in the domain of DataFlow computing. The volume is a collection of five chapters that were solicited from authorities in the field, each of whom brings to bear a unique perspective on the topic.

Chapter 1, "A New Course on R&D Project Management in Computer Science and Engineering: Subjects Taught, Rationales Behind, and Lessons Learned," by Milutinović *et al.* describes the essence of a course that was developed to complement a DataFlow course and to teach DataFlow researchers about issues of importance for promotion of their results with a commercial potential. The course prepares students for their professional life after graduation and especially it prepares them for the challenges related to efforts to bring new paradigm-shifting ideas into the commercial world. Authors attempted to justify each step in the process by at least one example.

In Chapter 2, "Advances in Dataflow Systems," Chau *et al.* present an overview of the parallel processor design and dataflow systems over the last 40 years. The article focuses on the renewed interest in dataflow systems caused by the advances in the domain of field-programmable technologies, and the use of field-programmable gate array chips to implement dataflow systems. Finally, this article discusses software design flow for dataflow systems as well as several high-performance computing applications implemented on dataflow systems.

In Chapter 3, "Adaptation and Evaluation of the Simplex Algorithm for a Data-Flow Architecture," Čibej and Mihelič articulate a novel adaptation of the classical simplex algorithm for a Data-Flow architecture. A detailed description of the linear programming problem and a few basic concepts, which impact the algorithmic complexity, are discussed. Moreover, it presents a novel and an efficient streaming Data-Flow implementation of the classical simplex algorithm on Maxeler Data-Flow system. This article shows significant speedup of the simplex algorithm implemented on a Data-Flow system over an optimized ControlFlow-based implementation.

In Chapter 4, "Simple Operations in Memory to Reduce Data Movement," Seshadri and Mutlu introduce two novel in-memory data processing techniques that improve the performance and energy efficiency by exploiting the underlying operations of the main memory technology in performing more complex tasks. The article first describes RowClone, a mechanism that exploits DRAM technology to perform bulk copy and initialization operations completely inside main memory. It then describes a complementary work that uses DRAM to perform bulk bitwise AND and OR operations inside main memory. The article introduces the idea of Processing using Memory (PuM), which exploits some of the peripheral structures already existing inside memory devices (with minimal changes), to perform other tasks on top of storing data. PuM is a cost-effective approach since it does not add significant logic structures near or inside memory.

Finally, in Chapter 5, "A Novel Infrastructure for Synergistic Dataflow Research, Development, Education, and Deployment: The Maxeler AppGallery Project," Trifunovic et al. present the essence and the details of a novel infrastructure, a dataflow AppGallery, that synergizes research, development, education, and deployment in the context of dataflow research. These new tools are not only tuned to the dataflow environment, but they are also tuned to synergize with each other, for the best possible performance in minimal time, counting from the moment when new researchers enter the dataflow arena, until the moment when they are able to deliver a quality code for maximal speed performance and minimal energy consumption. The effectiveness of the presented synergistic approach was measured empirically, using a group of students in a dataflow-oriented course at the School of Electrical Engineering, University of Belgrade. The results of the study are presented in this article.

We hope that readers find these articles of interest, and useful for teaching, research, and other professional activities. We welcome feedback on the volume, as well as suggestions for topics of future volumes.

ALI R. HURSON
Missouri University of Science and Technology,
Rolla, MO, United States
VELJKO MILUTINOVIĆ
University of Belgrade, Belgrade, Serbia

CHAPTER ONE

A New Course on R&D Project Management in Computer Science and Engineering: Subjects Taught, Rationales Behind, and Lessons Learned

Veljko Milutinović*, Stasa Vujicic Stankovic[†], Aleksandar Jovic[‡], Drazen Draskovic*, Marko Misic*, Danilo Furundzic[§]
*School of Electrical Engineering, University of Belgrade, Belgrade, Serbia
[†]School of Mathematics, University of Belgrade, Belgrade, Serbia
[‡]School of Physical Chemistry, University of Belgrade, Belgrade, Serbia
[§]School of Architecture, University of Belgrade, Belgrade, Serbia

Contents

Advances in Computers, Volume 106
ISSN 0065-2458
http://dx.doi.org/10.1016/bs.adcom.2017.04.001

1

Abstract

This chapter describes the essence of a course for senior level undergraduate students and for master students of computer science and engineering, and analyzes its effects. The course prepares students for their professional life after graduation, and especially, it prepares them for the challenges related to efforts to bring new paradigm-shifting ideas into the commercial world. This course was developed to complement a DataFlow course and to teach DataFlow researchers about issues of importance for promotion of their results with a commercial potential. Consequently, course examples and homework assignments were chosen to reflect issues of importance in the commercialization of the DataFlow concept. The course includes the following subjects (presented with DataFlow-related issues in mind): (a) Writing proposals for Research and Development in industry and academia, (b) Understanding the essence of the MBA/PhD degrees and preparing the GMAT/GRE analytical exam, (c) Understanding Capability Maturity Model Integration and learning how to write holistic strategic project plans, (d) Understanding Project Management and learning how to write detailed tactical project plans, (e) Writing business plans for venture capital or business angels, (f) Writing patent applications, (g) Writing survey papers for SCI (Science Citation Index) journals, (h) Writing research papers for SCI journals, (i) Making an Internet shop, (j) Making a MindGenomics campaign for the Internet shop, (k) DataMining from project history and project experiments, and (l) Preservation of project heritage and skills related to brand making. Each subject matter is covered by a homework assignment to help deepen the practical knowledge of the subject matter covered. In addition to the above described, which is accompanied with homework, the following four subjects are also covered and accompanied with in-class discussions (oriented to DataFlow research): (m) Inventivity, (n) Creativity, (o) Effectiveness, and (p) Efficiency. Consequently, the analysis part concentrates on the following issues: (a) Inventivity: How different majors react to the subject matters, (b) Creativity: How efficiently the initial knowledge gaps get bridged, (c) Effectiveness: How the experience of the teacher helps, and (d) Efficiency: How the previous experiences of students help.

ABBREVIATIONS

CE computer engineering
CMMI capability maturity model integration
CS computer science
GPA grade point average
HW homework
KI knowledge improvement

MG mind genomics
MI management informatics
NSF national science foundation
PH physical chemistry
PM project management
PTO patent and trademarking office
R&D research and development
SBA small business administration
SCI science citation index

1. INTRODUCTION

The main point behind this chapter is that, if one teaches a paradigm-changing subject or performs research in a paradigm-shifting domain, one also has to teach graduate students and research assistants all about entrepreneurial issues, so their novel paradigm-shifting ideas find the road to wide acceptance, more effectively. With the above in mind, a course was developed, to be offered in parallel with a DataFlow course, which teaches the skills needed in the process, beyond the scientific and engineering knowledge. That is the link between this DataFlow-oriented special issue and the education-oriented approach presented in this chapter. In other words, a new education approach is needed that helps paradigm shifts to become widely accepted. Since the course presented here is offered in conjunction with a DataFlow course, educational examples as well as homework assignments were chosen to help the DataFlow research become more successful.

The general structure of the course under consideration is depicted in Fig. 1. For similar courses, see Refs. [1,2]. For a discussion of issues relevant to this chapter, see Refs. [3,4]. Rationales behind the structure given in the Abstract are discussed next.

1.1 Before the Project Starts

(a) One has to learn how to provide funding (i.e., how to apply for Horizon 2020 projects in Europe or NSF (National Science Foundation) projects in United States; homework #1 is to write a proposal related to DataFlow technology, compatible with a challenge in the first forthcoming call of the chosen agency for financing the R&D work). (b) How to manage resources with a knowledge of the MBA or PhD degree (a possible homework #2 could be on how to pass an entrance exam for MBA or PhD degree studies oriented to DataFlow marketing or DataFlow research; an alternative for the homework #2 is to prepare a PowerPoint presentation to defend the

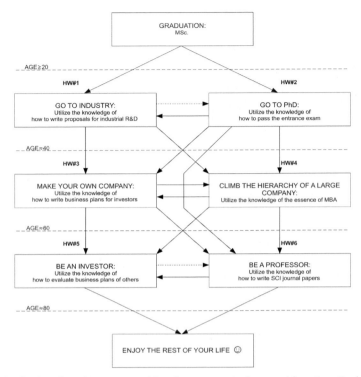

Fig. 1 Professional paths supported by the course under consideration. *Explanation*: One can take all paths indicated by *solid arrows* (those with *dotted arrows* are less likely to be taken); activities defined by *boxes* are most likely taken at the indicated age. It makes sense, at the age of 20+ that students get prepared for the later ages, including those at the age of 60+. *Legend*: HW, homework.

DataFlow technology-related proposal from homework #1 in front of the officers of the chosen agency for financing the R&D work).

1.2 Immediately After the Project Starts

(c) One has to know how to do strategic planning (e.g., CMMI, Capability Maturity Model Integration); homework #3 is related to the CMMI-based description of all related project issues, for the case that the afore-mentioned DataFlow proposal was accepted. (d) How to do a day-to-day planning of the DataFlow research based on the accepted proposal (e.g., Microsoft MS Project with a stress on agile planning); homework #4 is related to Microsoft MS Project-based description of specific core phases and insists on agile planning.

1.3 When the Project Is Close to Its End

(e) One has to write business plans for investors (and also has to know how investors look at the business plans of the applicants; homework #5 is to prepare business plan for forming a company based on the final product of the project defined by homework #1). (f) One has to know how to write a patent application (to protect the major product of his/her future company; homework #6 is to prepare a patent application related to the final product of the project defined by homework #1). Both above have to keep in mind specifics of DataFlow products and to stress these specifics, as well as their potential advantages on the market.

1.4 Soon After the Product Ends

(g) One has to write a paper of all possible solutions of the problem under consideration in the just finished project (homework #7 is to write a survey paper, to educate potential customers in the future). (h) One has to write a paper about the solution introduced in the project just finished (homework #8 is a research paper that explains the essence of the company product to potential company customers, which can help maximize company's sales; this chapter, if published in an science citation index (SCI) journal, can also help the author to obtain a PhD degree or can help the author's Alma Mater university to obtain a better ranking at world universities ranking lists). Both above have to be done a way that helps bring all the advantages of DataFlow products to the attention of its potential future users.

1.5 Soon After the Project Is Over

(i) One has to create an Internet shop (homework #9 is to establish an infrastructure for successful sales). (j) One has to implement a MindGenomics campaign (homework #10 is about maximizing the sales, using a sophisticated approach to personalized marketing). These two homework assignments should demonstrate that the students know how to make DataFlow issues work for their e-shop and the related MindGenomics campaign.

1.6 Long After the Project Is Over

(k) One has to know how to find hidden knowledge in the accumulated business or experimenting data (DataMining algorithms are taught and homework #11 is related to DataMining). (l) One has to know how to preserve the heritage of the entire project and to turn it to a brand (homework #12 is related either to heritage preservation or brand making). The goal of both assignments is to make students aware of the fact that hidden correlations of relevant issues do exist and could be helpful in branding.

1.7 An Introductory Remark

Topics covered are enlisted in Table 1. As indicated earlier, the course covers six different topics, each one with two subtopics, and each subtopic is accompanied by a homework assignment.

All above described knowledge is also of interest for the lifetime self-management of a computer engineer or scientist. Fig. 1 shows a number of possible routes that a student can take after graduation at the master level. The basic assumption here is that the master diploma will be obtained at the age of 20 or more and that a question is what happens next. Different stages of Fig. 1 require the knowledge described in Table 1.

In Serbia, this course was taught for several years now, as follows:

(a) In the School of Physical Chemistry, on the master level, as the only obligatory course on the master level, with 2h of teaching and 13h of homework per week. The course was also open to students in related fields. Majors of this type need the knowledge about DataFlow in order to substitute expensive reactors with cheap and fast simulations.

Table 1 List of Topics and Subtopics Covered by the Course Under Consideration in This chapter

Before the project starts:
(a) How to provide funding
(b) How to manage resources
Soon after the project starts:
(c) How to do strategic planning
(d) How to do day-to-day planning
Before the project ends:
(e) How to write business plans for investors
(f) How to write a patent application
Immediately after the project ends:
(g) How to write a survey paper
(h) How to write a research paper
Soon after the project ends:
(i) How to create an Internet shop
(j) How to implement a MindGenomics campaign
Long after the project ends:
(k) How to find hidden knowledge in data bases
(l) How to preserve the project heritage and how to brand it

Consequently, their homework assignments were to stress the values related to efficient and effective simulations.

(b) In the School of Mathematics, as a recommended elective course on the master level, for Computer Science and Informatics majors, with 2 h of teaching and 3 h of homework per week. The course was also open to students in related fields. Majors of this type need the knowledge about DataFlow in order to speed up real-time software using cheap and fast accelerators. Consequently, their homework assignments were to stress the values related to efficient and effective acceleration.

(c) In the School of Electrical Engineering, on the senior level, it is an elective course for Computer Engineering and Software Engineering majors, with 2 h of teaching and 3 h of homework per week. The course was also open to students in related fields. Majors of this type need the knowledge about DataFlow in order to learn how to create inexpensive accelerators. Consequently, their homework assignments were to stress the values related to efficient and effective hardware design and system software design.

(d) In the School of Business Administration, it has been offered periodically, on the PhD level, either for credit or not for credit. Majors of this type need the knowledge about DataFlow in order to learn how to use this new technology in online trading and risk assessment. Consequently, their homework assignments were to stress the values related to efficient and effective business applications.

Each school defines homework' focuses in the fields of interest for the school; however, in each particular case the emphasis is on some aspect of DataFlow technology, which is of interest for the particular School. For example, in the last academic year offerings of the course, in the School of Physical Chemistry, all homework assignments were related to energy conversion processes: in the School of Mathematics, to data base programming; in the School of Electrical Engineering, to Web and Mobile applications; and in the School of Business Administration, to business modeling. Of course, in each one of these cases, the major issue was how to create an effective DataFlow-based implementation.

2. PART 1: LEARN HOW TO CREATE A PROPOSAL FOR HORIZON 2020 OR NSF

In Europe, the Horizon 2020 (H20) initiative provides R&D funding for small and large businesses, as well as for small and large universities. Similar mechanisms exist in United States under NSF.

Students are first taught the rules of H20 (if the course is offered in United States, the NSF rules are taught). Then, they are asked to look into the currently open calls for proposals and to create an R&D idea compatible with one of the challenges of that call. Their ideas are discussed in front of the classroom. Students are told that it is OK if their ideas are naive or even wrong. They are told that now is the time to learn formalisms; 5 or 10 years later, they will be experienced professionals and will have great ideas, but will not be "afraid" of the formalisms.

The homework assignment is to create a H20 proposal (or an NSF equivalent), which is defended in front of the professor alone, and also to prepare a PowerPoint for presentation of the project, which has to be presented in front of the professor and all other students. Specific requirements for the PowerPoint include, as far as the essence, the following elements found also in comprehensive exams at the universities in the United States, on the MSc and PhD levels:

(a) Problem statement, existing solutions, and the proposed solution
(b) Axioms, conditions, and assumptions
(c) Evaluation, methodology in theory, simulation, and practice
And also, as far as the form:
(d) As few bullets as possible; the more animation—the better
(e) In multiple line bullets, one line—one thought (semantic brakes)
(f) On each slide, the current slide number and the total slide count (n/N)

3. PART 2: LEARN THE ESSENCE OF MBA AND PhD AND LEARN HOW TO PREPARE FOR THE GMAT/GRE ANALYTICAL EXAM

We consulted the Shanghai Top 500 List of the Best Universities (see Ref. [5]), and we have noticed that, out of the top 20, 16 are located in the United States, only 3 in EU, and only 1 in Japan.

Consequently, the stress in teaching, when it comes to the second educational topic, is how to pass entrance exam for the best universities in the United States, on the MBA and PhD levels. Basic premises of GRE/GMAT exams are taught (students are told that the examinee is treated as a computer with an input processor, the central processing unit, and the output processor) that the most critical element of these exams (analytical) is elaborated in full details, and a related homework assignment is presented.

The homework assignment is to take the GRE/GMAT analytical and to submit it for grading to the professor. The details about the GRE/GMAT can be found in Ref. [6].

4. PART 3: LEARN CMMI

Students are explained that project planning includes not only the core of the engineering part; also, a set of other nonengineering activities, all the way from the initial idea to the end of project life cycle.

Next, the essence of CMMI is taught, with an example from the school-related field, which is tuned to the specific needs of the school at which the course is taught. It was stressed to the students that CMMI approach was created by the Carnegie-Mellon Software Engineering Institute and was used in a number of applications, from software design to defense logistics.

The assumption here is that the H20/NSF project proposal was successful, that the project is awarded, and that a holistic plan has to be made for all relevant activities to follow till the end of the project.

The related homework is supposed to cover levels 1, 2, and 3 of the CMMI approach (the more detailed levels 4 and 5 are not covered). Details of each level have to be documented properly.

5. PART 4: LEARN PROJECT MANAGEMENT

Depending on the amount of teaching hours available, teaching the PM (Project Management) could be done by the professor or could be left to self-learning. Working with elements of software tools for PM (e.g., Microsoft MS Project) is considered important and should be learnt from manuals. In the classroom, the teacher should explain the essential differences of using the PM software in various fields. The emphasis is on agile methods. The homework #5 is set to cover a part of the "awarded" project from homework #1 for which the CMMI model was set in homework #4.

6. PART 5: LEARN HOW TO WRITE BUSINESS PLANS FOR INVESTORS

No matter which avenue one takes, the assumption is, sooner or later, the young professional will arrive at the age of about 40, and then he/she will be faced with a decision problem: He/she will notice that some much younger professionals are joining their company/workplace, and these young people will be much more skillful when it comes to the nuts and bolts of the new technologies. So, one has to think how and where to "escape" from the newcomers. This is true in general, in any organization. Students are told, the best place to "escape" is to the one of the management positions in the company or the research lab where they are, or elsewhere. If that is the decision, then there are two choices: (a) one is to start a new company and (b) the other one is to try to climb successfully the hierarchy of a large company. Of course, appropriate up-to-date information has to be delivered to students at this point.

If one decides to take the first path (start-up), then one has to learn how to write a business plan for investors. Therefore, this educational part includes a concise analysis of the ways in which one can do a business plan for both, venture capital and business angels.

Among the topics taught in this part of the course is the SBA methodology (Small Business Administration, [7]), which was developed at the Harvard University.

The homework #5 assignments is based on the assumption that the proposal in homework #1 resulted in a successful product, and the time has come to start a company around that product and to make money out of the research and development invested until that time.

7. PART 6: LEARN HOW TO PREPARE A PATENT APPLICATION

The major product of the newly formed company has to be protected. Consequently, students are taught the know-abouts of patenting. This course uses the methodology of the US PTO (Patent and Trademarking Office). The homework #6 is set to follow the guidelines from PTO.

8. PART 7: LEARN HOW TO WRITE SCI JOURNAL PAPERS OF THE SURVEY TYPE

In the educational part number 7, students are taught how to write survey papers for SCI journals. In the context of the newly formed company, such a paper can help in educating potential customers. In the context of doctoral research, students are explained that one cannot become a PhD unless one demonstrates an ability to publish in SCI journals and that the newest trend at universities all around the world is that one cannot get a PhD unless one publishes at least a 1, or 2, or even 3 SCI papers (the number of required SCI papers depends on the strength of the university and the realities of each field). For more information, see Scientific Thomson [8].

Students are explained that two types of SCI papers exist: survey papers and research papers. Survey papers are something that can be accomplished through a homework assignment in a one-semester course, and that is why homework #7 is to write a survey paper on a topic selected by the student.

If a student gets the highest grade on his/her homework (survey paper), he/she cannot get the points unless he/she shows a proof that the paper was submitted to a journal which accepts survey papers.

9. PART 8: LEARN HOW TO WRITE SCI JOURNAL PAPERS OF THE RESEARCH TYPE

A research paper is not that easy to generate and definitely not in one semester. Consequently, only a precise template for writing research papers is elaborated down to details, and students are asked to bring their previously concluded research rewritten according to the template just described. Students without previous research experience are given an opportunity, in a follow-up research-oriented course, to do a research project and to write a research paper from scratch.

In the course described here, one can only teach theoretical issues (see Refs. [9,10]), and one can encourage students, for extra homework credit, to write a research paper about their research conducted before the course started, using the formalism taught, step by step. Therefore, homework #8 is optional, for extra classroom credit.

10. PART 9: LEARN HOW TO MAKE AN eShop

If company sales are supported by an eShop, the revenues can increase significantly. Therefore, this course covers three different methods to create an eShop: (a) For small business (e.g., a YahooStore), (b) For medium business (e.g., ecBuilder), and (c) For large business (e.g., MS eCommerce Edition). The homework #9 asks students to create an eShop, to support the product of the company formed in homework #5, which is protected, as elaborated, by homework #6.

11. PART 10: LEARN MindGenomics

MG (MindGenomics) is taught as the superior approach to personalized marketing (see Ref. [11]). Students are explained the steps of the two-phase procedure and are given examples. The MG in the first phase determines different customer types (mostly three of them) and in the second phase determines the type of each given customer, without tracking the customer behavior, or asking the customer to fill forms. Homework #10 is to organize an MG campaign for the product being sold by the eShop developed in homework #9.

12. PART 11: LEARN BUSINESS INTELLIGENCE BASED ON DataMining

Ten different DataMining methods are taught, in the context of finding hidden knowledge, from the company database (in the commercial environment), or from scientific experiments (in the research environment). For each DataMining method, the math framework is given, as well as an anecdotic example from the World history. The homework #11 is to apply one of the 10 algorithms to a problem selected by each student individually.

13. PART 12: LEARN HOW TO PRESERVE HERITAGE AND HOW TO CREATE A BRAND

The essence of digital preservation of heritage is presented, and it is explained that heritage can refer not only to nations, states, cities, or families, but also to companies, projects, products, and inventions. Two different methods for creating a heritage portal are taught, and the essence of heritage mining is explained. Two different brand-making schools of thought are presented and discussed. The homework #12 is to contribute to a heritage preservation portal or to make a plan for branding.

14. ANALYSIS

As indicated earlier, the course has been taught for four different majors: CE (computer engineering, for the last 4 years), CS (computer science, for the last 3 years), MI (management informatics, for the last 2 years), and PH (physical chemistry, for the last 4 years). Table 2 shows results only for the first three majors, since the results for the fourth major were not easy to compare with the first three, due to intrinsic topic differences.

Table 2 Average Distribution of Grades of Students for Three Different Majors (CE, CS, MI), for Homework Assignments 1–6

CE	2	6	22	58	12
CS	1	7	72	12	8
MI	10	60	20	7	3
GRADES	>50	$\geq$60	$\geq$70	$\geq$80	$\geq$90

For example, 2 means in the horizontal CE line means that 2% of the students in the CE class got a grade larger than 50% and smaller than 60%
Explanation: Different majors perform differently on various homework assignments, which can be found on the website of the course, and is explained by their different initial aspirations at the time of major selection. The table was normalized so the class size did not matter (the class size was larger than 25 in all summarized cases). *Legend*: x = Grades of students on the scale from 51 to 100; y = Percentage {0%, 100%} of students that received the given grade.

Students of various majors enter the course with various backgrounds, and consequently, they score differently on homework assignments. Table 2 gives the distributions of scores for the three majors, for all required homework assignments. From Table 2 it is that the distributions of grades were different for different majors (58% of students had a grade equal or higher than 80, for the CE major; 72% had a grade equal or higher than 70, for the CS major; 60% had a grade equal or higher than 60, for the MI major). A more detailed analysis concludes, as expected, that CE majors did best on the writing of proposals and papers, MI majors did best on making and evaluating of business plans, while the CS majors did best on Web and analytical assignments.

An appropriate pretest was given to all students at the entry point, and the final test at the exit point. All three majors scored about the same at the final exam, which means that students at the age of Master bridge easily the knowledge gaps. Fig. 2 shows the pretest to final test knowledge improvements for the analyzed majors. The initial expectation was opposite from what Fig. 2 shows; this could be attributed to different motivation factors.

Fig. 3 demonstrates that students keep improving from one generation to the other (probably mostly due to the improvement in presentation skills of the professor), and also that it takes less and less time to cover the material (probably mostly due to the better preknowledge gained at university or around university).

Table 3 summarizes some typical comments of those students who contacted their professor 1 or more years after graduation. These comments shed light on the differences between the professional life realities and the classroom-time expectations at the time of studying.

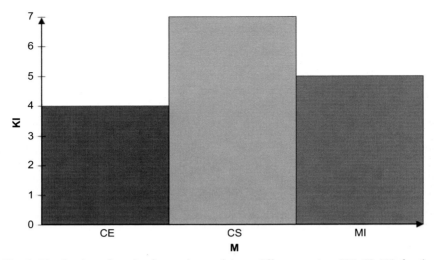

Fig. 2 Distribution of grades for students of three different majors (CE, CS, MI), for the final exam relatively to the pretest. *Explanation*: The histogram reflects the level of pre-knowledge that various majors bring into the classroom. *Legend*: *KI*, knowledge improvement (grade at the final test divided by the grade at the pretest); *M*, majors (CE, CS, or MI).

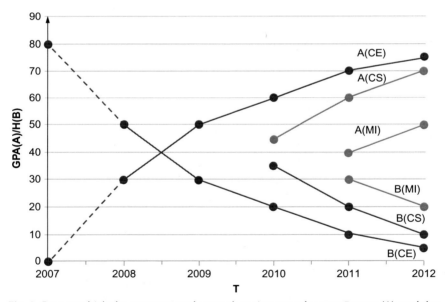

Fig. 3 Rate at which the average grade was changing over the past 5 years (A), and the number of hours needed to cover the subject matters (B). *Explanation*: From generation to generation, grades get better (because it took the time to the teacher to develop good case studies), while it took less and less time to teach (because students also learn from media and peers); this left more and more time for in-class student presentations. Note that the GPA increased as course was covered in shorter amount of time; this is due to the fact that initial knowledge of students at the course entry point was increases as the course went by. *Legend*: *T*, time; *GPA*, grade point average; *H*, hours needed to teach the subject. The *dotted line* refers to experimental teaching, prior to inclusion of the course into the official curriculum.

Table 3 Typical Comments of Those Who Return Their Impressions After Some Years in Industry or Academia

Homework 1	I felt this homework was not useful for me, but suddenly I was faced with the challenge and had to write an H2020 proposal for my boss
Homework 2	I felt this homework was not useful for me, since I did not plan to go that way in my life, but I see that knowledge will be useful to me to guide my children
Homework 3	The CMMI taught me to be holistic not only in profession but also in life too
Homework 4	It was useful to learn project management realities. Saved me some time afterward
Homework 5	I always dreamed to have my own startup, and I knew that this homework will be useful to me, which it was
Homework 6	Writing patent applications is fun, but we do not know how useful it is to invest effort on patenting algorithmic ideas
Homework 7	I felt I would never write a scientific paper, but now I see, I keep doing it
Homework 8	When writing the research paper, we get to understand our own work better!
Homework 9	I felt I was not able to sell my products, services, ideas, but now I know I can
Homework 10	MindGenomics is an invention of a genius!
Homework 11	Data Mining is best understood inside the framework of a real project
Homework 12	Heritage sites are important and useful, but heritage mining is a real challenge

In the last 4 weeks of the course, students are taught how to be efficient [12], effective [13], inventive [14], and creative [15]. Initially, these four topics were not taught. After the four topics were added to the course, the quality of homework assignments went up. The Refs. [12–15] were studied in the context of the Maxeler Technologies [16], a provider of DataFlow systems.

15. CONCLUSION

15.1 What Was Achieved?

A course is generated to motivate students to think about holistic project management with a stress on DataFlow research and also to think about their short- and long-distance professional and life trajectories. The presented course gives them both theoretical and practical knowledge about activities of interest at various stages of a R&D project, in general and specifically for the case of DataFlow research; the presented practical knowledge is also useful for their general professional development.

15.2 Who Will Benefit From the Course?

This course is of benefit not only to students and their teachers but also to university committees engaged in planning their curricula. The subject is of interest not only for computer engineering students and educators but also for those in number of scientific areas.

15.3 Newly Open Problems?

An important problem for future is how to extend this course beyond one semester, so the homework assignments can be more mature. The more professional the homework assignments, the higher the benefit for students. Another possible problem is how to make it better suited for in-house industrial courses.

15.4 What Is to Remember About This Course?

Students are asked to call their professor 5, 10, or 20 years after graduation, to share their experiences with him/her, so the course could be made better (for those interested to teach it in the future). Students' feedback is crucial for this type of courses.

ACKNOWLEDGMENTS

This research was conducted through the projects 44006 and III 47003 financed by the Serbian Ministry of Science. The authors are thankful to Marija Jovic from the University of Belgrade for her help related to teaching of the subject matter discussed in this chapter. Also, to all those who helped when this course was conducted, in short forms in various regions—at the universities in the United States (Dartmouth, Harvard, NYU, CUNY Albany), Japan (Tokyo and Sendai), and Europe (Barcelona, Valencia, Salerno, Siena, Karlskrona, and Shoevde).

REFERENCES

[1] B. Burnett, D. Evans, Designing Your Life, http://www.stanford.edu/class/me104b/cgi-bin/, 2011.
[2] D. Patterson, How to Have a Bad Career in Research/Academia, https://people.eecs.berkeley.edu/~pattrsn/talks/badcareer.pdf, 2001.
[3] V. Milutinovic, Wisdom of education for globalization, IPSI Trans. Internet Res. 1 (1) (2005) 3–5.

[4] V. Milutinovic, S. Tomazic, How to ruin the career of a PhD student in computer science and engineering: Precise guidelines, IPSI Trans. Internet Res. 4 (2) (2008) 24–26.

[5] The 2016 Academic Ranking of World Universities. http://www.arwu.org. 2016.

[6] The GRE and GMAT Tests. http://www.800score.com. 2016.

[7] The US Small Business Administration, SBA. http://www.sba.gov. 2011.

[8] Scientific Thomson. http://scientific.thomson.com. 2009.

[9] V. Milutinovic, The best method for presentation of research results, The TCCA Newsletter (1995) 1–6.

[10] V. Milutinovic, A good method to prepare and use transparencies for research presentations, The TCCA Newsletter (1996) 1–6.

[11] H. Moskowitz, A. Gofman, Selling Blue Elephants, Prentice Hall, Upper Saddle River, NJ, USA, 2007.

[12] K.L. Lloyd, G.B. Anderson, Guide to Selling, Dorling Kindersley Publishing, UK, 2001.

[13] P.F. Drucker, The effective executive, Harper Business, NY, USA, 1966.

[14] M. Perl, Getting good ideas in science and engineering, IPSI Trans. Adv. Res. 3 (2) (2007) 3–7.

[15] J. Friedman, Why we need basic research, IPSI Trans. Adv. Res. 4 (2) (2008) 2–5.

[16] M. Flynn, O. Mencer, V. Milutinovic, G. Rakocevic, P. Stenstrom, R. Trobec, et al., Moving from petaflops to petadata, Commun. ACM 56 (5) (2013) 39–43.

ABOUT THE AUTHORS

Veljko Milutinović
Life Member of the ACM
Fellow Member of the IEEE
Member of Academia Europaea
Member of the Serbian Academy of Engineering
Member of the Advisory Board of the Vienna Congress COMSULT
Member of the Scientific Advisory Board of Maxeler Technologies

Prof. Veljko Milutinović (1951) received his PhD from the University of Belgrade, spent about a decade on various faculty positions in the USA (mostly at Purdue University), and was a codesigner of the DARPAs first GaAs RISC microprocessor. Later, for almost three decades, he taught and conducted research at the University of Belgrade, in EE, BA, MATH, and PHYS/CHEM. Now he serves as the Chairman of the Board for the Maxeler operation in Belgrade, Serbia. His research is mostly in datamining algorithms and dataflow computing, with the emphasis on mapping of data analytics algorithms onto fast energy efficient architectures. For seven of his books, forewords were written by seven different Nobel Laureates with whom he cooperated on his past industry sponsored projects. He has over 40 IEEE journal papers, over 40 papers in other SCI journals (4 in ACM journals), over 400 Thomson-Reuters citations, and about 4000 Google Scholar citations. Short courses on the subject he delivered so far in a number of universities worldwide: MIT, Harvard, Boston, NEU, Columbia, NYU, Princeton, Temple, Purdue, IU, UIUC, Michigan, EPFL, ETH, Karlsruhe, Heidelberg, etc. Also at the World Bank in Washington DC, BNL, IBM TJ Watson, Yahoo NY, ABB Zurich, Oracle Zurich, etc.

Stasa Vujicic Stankovic, PhD, is a teaching assistant at the University of Belgrade, School of Mathematics, Serbia. She holds PhD (2016) in Computer Science and Informatics from the same university. She teaches several courses on programming, algorithms and data structures, information systems, and research and development project management in computer science. Her research fields include natural language processing, and in particular information extraction, information retrieval and web search, formal languages and automata theory, databases, data minig, semantic web, as well as computer science education and educational tools.

Aleksandar Jovic, PhD student, is a teaching assistant at the University of Belgrade, School of Physical Chemistry, Serbia. He holds MSc (2008) in Physical Chemistry from the same university. He teaches several courses on physical chemistry, chemical thermodynamics, chemical kinetics, colloids, and physicochemical methods and research methodology. His research interests are mainly focused on chemical kinetics and catalysis, zeolites, electrode materials based on zeolites, electrochemical detection of pollutants, but also higher education, education tools, and research methodology.

Drazen Draskovic, PhD student, is a teaching assistant at the University of Belgrade, School of Electrical Engineering, Serbia. He holds BSc (2009) and MSc (2011) in Software Engineering from the same university. He teaches several courses on software testing, software engineering, computer networks, internet programming, and intelligent systems. His research fields include expert systems and artificial intelligence, software engineering, software testing, and education.

Marko Misic, PhD student, is a teaching assistant at the University of Belgrade, School of Electrical Engineering, Serbia. He holds BSc (2007) and MSc (2010) in Computer Engineering from the same university. He teaches several courses on programming, algorithms and data structures, parallel programming, and multiprocessor systems. His research interests are mainly focused on parallel and distributed programming with special attention to GPU programming, but also engineering education and educational tools.

Danilo Furundzic, PhD, is a teaching assistant at the University of Belgrade, School of Architecture, Serbia. He holds Dipl.-Ing. (2004) and PhD (2017) in Architecture and Urbanism from the same university, and MS (2006) in Industrial Engineering from l'École Centrale Paris. He teaches several courses on urban planning, infrastructure and management. His research interests include real-estate investment process modeling, urban parameters profitability evaluation, planning decisions feasibility checking, and education for sustainable development.

CHAPTER TWO

Advances in Dataflow Systems

Thomas Chau*, Pavel Burovskiy†, Michael Flynn‡, Wayne Luk§

*Intel Corporation, High Wycombe, United Kingdom
†Maxeler Technologies, London, United Kingdom
‡Stanford University, Stanford, CA, United States
§Imperial College London, London, United Kingdom

Contents

Abstract

Research on processors based on dataflow principles has been active for over 40 years, but such processors have not become mainstream general-purpose computing. Recently, advances in field-programmable technology spark renewed interest in

Advances in Computers, Volume 106
ISSN 0065-2458
http://dx.doi.org/10.1016/bs.adcom.2017.04.002

dataflow systems. This chapter first provides an overview of parallel processor design and dataflow systems. Then the use of reconfigurable chip such as field-programmable gate array to implement dataflow machines is discussed, followed by an introduction of software design flow for dataflow hardware. After that, several high-performance computing applications developed using dataflow machines are explored.

ABBREVIATIONS

CAPI coherent accelerator processor interface
CSR compressed sparse row
DFE dataflow engine
DSP digital signal processor
FEM finite element method
FPGA field-programmable gate array
HPC high-performance computing
PaaS platform-as-a-service
PEs processing elements
PQ proximity query
SIMD single instruction multiple data
SLiC simple live CPU interface
SMC sequential Monte Carlo
SpMV sparse matrix–vector multiply
VLIW very large instruction word

1. INTRODUCTION

The design of dataflow processors was pioneered around the mid-70s. In contract to conventional Von-Neumann model of computation which program instructions are executed sequentially in a specific order, dataflow processors focus on optimizing the movement of data in applications and utilize massive parallelism to improve performance. Despite much research over the last few decades as discussed in Refs. [1–3], such dataflow processors have not become mainstream general-purpose computing. This situation, however, can be changing. Recent advances in field-programmable gate arrays (FPGAs) allow computer chips to be reprogrammed after manufacturing and at runtime, therefore, enable dataflow systems to be customized for various applications. FPGAs combine an abundance of logics gates, registers, digital signal processors (DSPs), memory, and I/Os, which led to novel dataflow systems that are suitable for massive parallelism and have significant benefits in performance and in energy efficiency. Dataflow machines have attracted interest of the hyperscale companies in data center, for instance, Microsoft's Project Catapult [4] uses dataflow machine pervasively to accelerate their search engine and cloud service, Baidu is using dataflow machines to accelerate machine learning applications in their data centers [5], and Amazon has brought dataflow machines to their cloud service [6].

This chapter provides an overview of dataflow systems. We begin with a discussion of parallel processor design in Section 2 and how the approach of parallel computing shifts from multiprocessor software to dataflow computing in Section 3. Then Section 4 introduces some state-of-the-art dataflow machines implemented by reconfigurable technologies. In Section 5, a software design flow for dataflow machines is described: computation is defined by dataflow graphs such that the data flows through a spatial system composed by arithmetic operations. This is in contrast to traditional software programming model which assumes a program computes in time. Finally, Sections 6–9 describe recent appearances of the dataflow systems in many high-performance applications.

2. PARALLEL PROCESSOR DESIGN

Parallel processors have been an integral part of computer architecture and design for more than 50 years. Indeed, it is almost 50 years since the great debate about the future of parallel processing in computing occurred. The debate occurred in 1967. Two well-known computer engineers of the day—Gene Amdahl of IBM spoke for the serial processing approach, while Daniel Slotnick, a Professor at the University of Illinois and the chief architect of the Iliac IV, spoke for the parallel approach.

Amdahl [7] made two points. The first was an observation that serial processes are inherently better because they avoid the extensive time spent programming parallel processors. He coupled this with the note that technology regularly improves the performance of serial processor. This could be seen as a variation of Moore's law on frequency scaling. Amdahl's second point was an observation that we now know of as Amdahl's law. For a parallel machine containing p processors with s being the fraction of serial code and T_1 the time to execute the program on serial machine, the speedup Sp over the serial machine is limited to:

$$Sp = \frac{T}{T \cdot s + T \cdot (1 - s)/p} \qquad (1)$$

or

$$Sp = \frac{1}{s + (1-s)/p} \tag{2}$$

Slotnick [8] acknowledged programming difficulties in dealing with parallel processors. He writes:

> The parallel approach to computing does require that some original thinking be done about numerical analysis and data management in order to secure efficient use. In an environment which has represented the absence of the need to think as the highest virtue this is a decided disadvantage.

But he believed that ultimately the parallel approach would win out technologically because it is based on replication.

More recently some observers have labeled his argument as Slotnick's law of effort:

> ... Speedup is achieved by algorithmic, analytic & programming effort ...

In the ensuing 40 years, Amdahl's argument prevailed as most computing engineering effort went into serial processors. But about 10 years ago frequency scaling ceased due to power density limitations.

As a general rule higher frequency processor designs require exponentially more power than lower frequency designs. The standard relationship between processor execution time, T, and required power, P, is:

$$T^3 P = k$$

where k is a constant [9]. As an example, if a high frequency design was designed to double the frequency of a particular reference design, one would expect it to consume eight times the power. While cleverer design may avoid the worst consequences of this, it is obvious that parallel designs will have a uniform power density that scales with area. The complexity of lower frequency parallel design coupled with reduction in off-chip traffic resulting from dataflow will bring a huge performance advantage.

With the continuing advance in circuit density, replicating processing units in the single instruction multiple data (SIMD) or multiple instruction multiple data configuration has been proposed. Various other forms of parallel processors, such as superscalar, very large instruction word (VLIW), array-based and vector-based processing, have also been developed.

While the parallel approach promises to offer maximum potential performance, realizing anything close to maximum performance has been a significant challenge.

3. FROM MULTIPROCESSOR SOFTWARE TO DATAFLOW COMPUTING

The conventional approach to programming high–performance computing (HPC) applications is to use some languages such as C++ with various augmentations to implement an application on as many processors as possible, up to a point of node capacity or speedup saturation. This approach has advantages of hiding implementation complexity and improving programmer productivity.

However, such approach has performance limitation due to enforced sequencing. Operations that can be executed in parallel are often serialized. To bring used data close to the processor and predict the control flow, large caches and very complex control are created. Multiprocessor is difficult to scale-up while meeting cost, power, and reliability requirements. Even for the so-called embarrassingly parallel HPC applications, traditional multiprocessor is difficult to get linear speedup with increasing numbers of processors.

Technology provides steadily growing number of transistors per unit area. Instead of computing in time, computation can be performed in dataflow/spatial systems, where the data flows through a structure composed by arithmetic operations laid out on the chip surface. Dataflow systems are based upon the notion that a computational action can execute as soon as the requisite operands are present at the inputs to its functional unit. The result is then forwarded to the next computational action. Dataflow computation avoids the complex control structures as seen in control flow computing, such as multilevel caches, instruction decode logic, branch prediction, and out-of-order scheduling.

Dataflow computation is defined by the dataflow graph, an essential part of dataflow computation. Dataflow computation was proposed by Dennis [10], and his colleague, Arvind [11], did much to broaden the applicability of dataflow computing. In the intervening years, dataflow has become the basis of computational models, programming languages, and hardware implementation. In Ref. [10], a processor was proposed to execute dataflow program expressed in a Fortran-like language. The processor has a unique architecture to avoid processor switching and memory/processor interconnection that usually limit the degree of concurrent processing. In Ref. [12], Treleaven et al. identified the concepts and relationships that exist within data-driven architecture. They examined data-driven architecture and developed models of computation, stored program representation, and machine organization. In Ref. [13], the Manchester project designed a dataflow processing engine based on dynamic tagging. In Ref. [14], the MIT tagged-token dataflow project used a high-level language with fine-grained parallelism and determinacy implicit in its operational semantics. Programs were compiled to dynamic dataflow graphs and executed on a tagged-token dataflow architecture. In Ref. [15], a restricted dataflow architecture was proposed which only executes a small subset of the entire program in the dataflow engine (DFE) at any one time to exploit local parallelism rather than global parallelism. EDGE architecture [16] is another restricted dataflow architecture being proposed to use control flow between dataflow blocks. It supports conventional memory semantics within and across blocks, permitting them to run traditional imperative languages such as C or Java, while gaining many of the benefits of more traditional dataflow architectures. Historical surveys are available, such as Refs. [1–3].

Dataflow computing requires a custom chip for a specific application. It is generally impractical to have machines that are completely optimized for only one code. Reconfigurable chip, FPGA in particular, can be reprogrammed at runtime to implement different applications, or different versions of the same application. For example, the coherent accelerator processor interface (CAPI) on IBM POWER8 systems [17] enables a custom acceleration engine to be connected to the coherent fabric of the POWER8 chip. CAPI allows dataflow computing by mapping application-specific, computation-intensive algorithms on an FPGA for acceleration. CAPI removes the overhead and complexity of the I/O subsystem, allowing the acceleration engine to operate with a smaller programming investment. FPGAs' reconfigurability allows hardware to be specialized without the traditional costs of hardware fabrication.

The Manchester project [13,18] has constructed an operational tagged-token dataflow processor to tackle realistic applications. The basic structure is a ring of four modules connected to a host system via an I/O switch module. The modules operate independently in a pipelined fashion. Tokens are encapsulated in data packets that circulate around the ring. Token packets destined for the same instruction are paired together in the matching unit. A wide variety of parallel programs are accelerated by the active function units in a single-ring system. Sl/Sinf measures program parallelism to indicate the suitability of a program for the dataflow architecture, where S1 is the total number of instructions executed and Sinf is the total number of simulated time steps required. The effectiveness of the software system has been improved by studying the ratio of instructions executed for each useful floating-point operation.

OpenCL is an industry standard parallel language based on "C." It has been applied to FPGA by vendors such as Intel to program FPGAs at a level of abstraction closer to traditional software-centric approaches [19]. Designers are able to take advantage of the capabilities offered by FPGAs, while using a high-level design entry language that is familiar to software programmers. OpenCL tries to address the challenge of mapping a "C"-based language to dataflow systems. C language has implicit assumptions that the underlying architecture executing these programs is a processor-based architecture, characterized by a sequence of instructions that control a datapath that manipulates data values stored in a memory. Conversely, FPGA architectures are more suited to implementing dataflow computing.

This chapter focuses on more recent hardware implementations of dataflow computation especially that done by Maxeler Technologies [20]. This approach uses FPGAs to implement and to emulate dataflow computation by creating executable version of the dataflow graph of an application.

4. DATAFLOW HARDWARE

Dataflow graphs are basic to any dataflow implementation. These graphs can be synchronous or asynchronous, static or dynamic, with a number of other variations. To simplify programming analysis and implementation the Maxeler approach uses synchronous, static dataflow graphs. The dataflow graph is emulated by a large FPGA implemented on a PCI express card, which is also supported by up to 96 GB high-speed buffer memory. An application can be implemented completely as a dataflow graph on dataflow machines or for very large applications only the kernels are implemented as dataflow machines. This limits the effects of Slotnick's law to a small portion of the application. The FPGA array is used to stream data results through a long computational pipeline keeping intermediate results either in the array or in an associated buffer. In principle, this eliminates the memory bottleneck associated with many HPC applications. Intermediate results are not returned to memory, so memory only has to provide the initial data and store the final results.

Fig. 1 shows an implementation where the applications or kernels are implemented as dataflow machines with multiple channels of high-speed

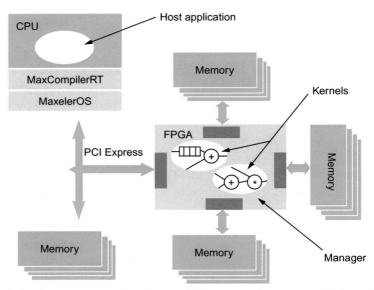

Fig. 1 A dataflow system emulated by an FPGA on a dataflow engine (DFE) card, interfacing a host processor memory with PCI Express.

memory directly coupled to the dataflow array. The PCI express card with the dataflow machine (emulated by the FPGA), the memory, and the memory controllers is referred to as the DFE.

System designers develop applications for dataflow machines using static synchronous streaming implementation. Generally, the design goal is throughput, so techniques such as loop unrolling are employed. The dataflow machine is fully synchronous and especially synchronized to the memory. Dataflow speedup is achieved by extensive pipelining in the dataflow graph. New data are brought into the dataflow chain each cycle, and the time through the dataflow machine is always the same. Current FPGA technology enables of the order of 10,000 nodes per device. If enough silicon is available, multiple dataflow graphs can be implemented on a single DFE. In addition, applications can be spread across multiple DFEs [21].

Maxeler [20] provides a range of dataflow computing platforms for different types of application:

- MaxWorkstation is a microATX form factor system which is equipped with one Vectis DFE, which has a Xilinx Virtex-6 XC6VSX475T FPGA (297,600 LUTs, 595,200 registers, 2016 DSPs, 38,304 kb of memory blocks). The DFE can have up to 48 GB DDR3 DRAM, and it is connected to an Intel i7-870 CPU (4 physical cores, 8 threads in total, clocked at 2.93 GHz) via a PCI Express Gen2 x8 bus.

- MPC-C500 is a 1 U server accommodating four Vectis DFEs. Each DFE can have up to 48 GB DDR3 DRAM and is connected to the CPUs via PCI Express, and DFEs within the same node are directly connected with MaxRing interconnect.

- MPC-X2000 is a 1 U server accommodating eight Maia DFEs. The FPGA part is Intel Stratix V GS 5SGSD8 (262,400 ALMs, 1,050,000 registers, 3926 DSPs, 50,000 kb of memory blocks). Each DFE provides up to 96 GB DDR3 RAM. The MPC-X series enables remote access to DFEs by providing dual FDR/QDR InfiniBand connectivity and direct transfers from CPU node memory to remote DFEs without inefficient memory copies.

5. DATAFLOW SOFTWARE

Fig. 2 illustrates the Maxeler design flow which converts dataflow graphs to hardware. Designs that target DFEs are captured using the Java-like

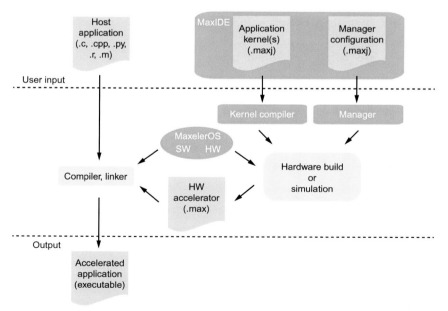

Fig. 2 Maxeler design flow from dataflow graphs to hardware.

MaxJ language and are compiled using kernel compiler. A design which consists of a number of locally synchronous kernels is connected together and to other asynchronous DFE resources using the manager. The kernel and manager are built to hardware configuration using FPGA vendor tools such as Intel Quartus Prime [22] and Xilinx Vivado [23]. The hardware configuration is loaded onto a DFE which communicates with the host software using a automatically generated API accessible from a variety of languages such as C, R, and Python.

A kernel describes computation and data manipulation in the form of dataflow graphs. MaxCompiler transforms kernels into fully pipelined synchronous circuits. Fig. 3 shows how a simple MaxJ kernel maps to a dataflow graph. The circuits are scheduled automatically to allow them to operate at high frequencies and make optimal use of the resources on the DFE. MaxCompiler supports multiple kernels for control flow and dataflow for maximum exploitation of low-level parallelization. This architecture allows the clock rate for each kernel to be individually tweaked to provide the optimum data throughput.

A manager describes the asynchronous parts of a DFE configuration which are connected together using a simple point-to-point interconnection scheme. The manager allows developers to connect resources such as memory and PCI Express streams to kernels to keep them fed with data. DFEs are controlled from a CPU using the simple live CPU interface (SLiC) API.

Tools are being developed using the Maxeler design flow with enhanced run-time reconfiguration and self-aware tuning. In Ref. [24], a design approach is proposed to automatically identify and exploit run-time reconfiguration opportunities while optimizing resource utilization. A hierarchical graph structure named as reconfiguration data flow graph is proposed to synthesize reconfigurable designs. Function analysis identifies reconfiguration opportunities through function property extraction and data dependency assignment. Available hardware resources are exploited dynamically by optimizing generated configurations based on function properties. Run-time solutions are generated by grouping configurations in different time slots. An ending-edge search algorithm is proposed to reduce the search space by introducing hardware design rules. In Ref. [25], a self-aware tuning and evaluation method is proposed for finite-difference applications targeting dataflow systems. Based on algorithm characteristics, design properties, available resources, and run-time data

A

```
class MovingAverageKernel extends Kernel
{
    MovingAverageKernel(KernelParameters parameters)
    {
        super(parameters);

        DFEVar x = io.input("x", dfeFloat(8, 24));

        DFEVar prev = stream.offset(x, -1);
        DFEVar next = stream.offset(x, +1);
        DFEVar sum = prev + x + next;
        DFEVar result = sum / 3;

        io.output("y", result, dfeFloat(8, 24));
    }
}
```

B

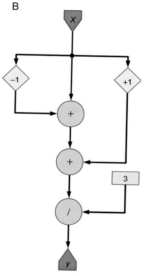

Fig. 3 MaxJ source code for (A) a simple moving average kernel and (B) the resulting generated dataflow graph.

size, the design process is aware of tuning opportunities to automatically capture optimized design with maximum performance. At compile time, awareness in the design process refers to proper estimation of resource consumption for possible designs, which enables exploration of design space without going through time-consuming synthesis tool. At runtime, awareness of benefits and overhead of generated designs enable quick evaluation of run-time performance.

6. DATAFLOW SYSTEM IN CLOUD COMPUTING

Next-generation cloud platform incorporates heterogeneous technologies such as DFEs, programmable routers, and different types of storage devices. An enhanced cloud platform-as-a-service (PaaS) software stack that makes effective use of DFEs can provide increased performance, reduced energy consumption, and lower cost profiles. HARNESS project [26] aims to develop a cloud platform that fully utilizes heterogeneous computing resources. The platform consists of a computation management system that allows heterogeneous computing resources to be allocated at runtime to satisfy a given goal or policy. Resource virtualization enables multiple processes share a single DFE. A virtual DFE can be instantiated to employ a fixed or variable number of physical DFEs. As physical DFEs are automatically allocated from one virtual resource to another depending on the workload, cloud providers are able to maximize resource utilization. In addition, a single software language (FAST [27]) is used to implement applications that target different types of computing resources. At the same time, domain-specific knowledge is codified using LARA [28] to derive designs that are optimized for a specific cloud platform and infrastructure.

Under the context of the HARNESS project, a run-time system with DFEs that supports elastic management is developed [29]. Elasticity is the ability to provision and release DFE resources at runtime based on computation demands, providing cloud users with the illusion of unlimited resources, and enable resources being shared across multiple applications. The run-time system allocates the right amount of resources to each application to satisfy its quality of service requirements while maximizing resource utilization and profit. An elastic management system for DFEs is developed to operate on a multiuser cloud environment. As shown in Fig. 4, it consists of an elasticity manager which decides an execution schedule to minimize the objective function and to adjust the resource pool size, and a resource manager which allocates jobs to resources based on schedule. An implementation library associates performance metrics with dataflow designs, and various scheduling strategies are used with the management system to achieve load balancing, reduce average waiting, and realize elasticity.

SHEPARD [30] is another framework under HARNESS that enables cloud platform to incorporate heterogeneous computing devices with DFEs and has the ability to decide at runtime the optimal compute device to execute a task. Application developers replace static task allocation such as

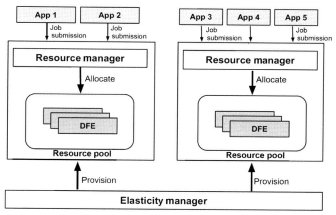

Fig. 4 Elastic management system for DFEs in a multiuser cloud environment. *Reprinted from P. Grigoras, M. Tottenham, X. Niu, J.G.F. Coutinho, W. Luk, "Elastic management of reconfigurable accelerators, " in: Proceedings of International Symposium on Parallel and Distributed Processing with Applications, pp. 174–181, 2014, © 2014 IEEE.*

function calls with SHEPARD managed tasks, which are managed at runtime using information stored in a database to determine which compute device is the most suitable to execute a task. Experiment results show that on a database management function that optimizes data storage, SHEPARD achieves good speedup over round-robin allocation and matches a static allocation by data type which requires prior knowledge of the data set.

Another resource management approach for heterogeneous architectures with DFEs has been proposed [31]. Both offline learning processes and online monitoring are used such that performance can be estimated from past observations during execution. The resource management system considers data locality and transfers costs to improve scheduling efficiency.

7. EXAMPLE APPLICATION ONE: SEQUENTIAL MONTE CARLO SYSTEM AND ITS GENERATION[a]

Sequential Monte Carlo (SMC) methods, also known as particle filter, are a set of posterior density estimation algorithms that perform inference of unknown quantities of interest from observations [33]. The observations arrive sequentially in time, and the inference is performed

---

[a] Reprinted from Ref. [32], © 2014 IEEE

online. SMC methods are often preferable to Kalman filters and hidden Markov models, as they do not require exact analytical expressions to compute the evolving sequence of posterior distributions. SMC methods work well for dynamic systems involving nonlinear and non-Gaussian properties, and they can model high-dimensional data using nonlinear dynamics and constraints, are parallelizable, and can greatly benefit from hardware acceleration. SMC has been studied in various application areas including object tracking [33,34], robot localization [35,36], speech recognition [37], and air traffic management [38–41]. For these applications, it is critical that high sampling rates can be handled in real-time. SMC methods also have applications in economics and finance [42] where minimizing latency is crucial.

SMC keeps track of a large number of particles, each of which contains information about how a system would evolve. The underlying concept is to approximate a sequence of states by a collection of particles. Each particle is weighted to reflect the quality of an approximation. One drawback of SMC is its long execution times so its practical use is limited.

In SMC, the target posterior density $p(s_t | m_t)$ is represented by a set of particles, where s_t is the state and m_t is the observation at time-step t. A sequential importance resampling algorithm [43] is used to obtain a weighted set of N_p particles $\{s_t^{(i)}, w^{(i)}\}_{i=1}^{N_p}$. The importance weights $\{w^{(i)}\}_{i=1}^{N_p}$ are approximations to the relative posterior probabilities of the particles such that $\sum_{i=1}^{N_p} w_t^{(i)} = 1$. This process is described in Algorithm 1. A more detailed description can be found in Ref. [33].

1. Initialization: Weights $\{w^{(i)}\}_{i=1}^{N_p}$ are set to the same value, e.g., $\dfrac{1}{N_p}$.

2. Sampling: Next states $\{s'_{t+1}^{(i)}\}_{i=1}^{N_p}$ are computed based on the current state $\{s_t^{(i)}\}_{i=1}^{N_p}$. The states can be simulated forward over the prediction horizon for H sampling intervals.

3. Importance weighting: Weight $\{w^{(i)}\}_{i=1}^{N_p}$ is updated based on a score function which accounts for the likelihood of particles fitting the observation. Within each iteration of itl_outer, the sampling and importance weighting stages are iterated itl_inner times so that those particles with sustained benefits are assigned higher weights. itl_inner increases as a function of idx1, because a larger idx1 implies that the set of particles reflects a more accurate approximation.

4. Resampling: Particles with small weights are removed, and those with large weights are replicated. This process is repeated for itl_outer times

Algorithm 1 SMC methods.

 1: **for** each time-step t **do**
 2:     $idx1 \leftarrow 0$
 3:     Initialisation
 4:     **while** $idx1 \leq itl\_outer$ **do**
 5:         $idx2 \leftarrow 0$
 6:         $itl\_inner \leftarrow f(idx1)$
 7:         **for** each particle p **do**
 8:             **while** $idx2 \leq itl\_inner$ **do**
 9:                 Sampling
10:                 Importance weighting
11:                 $idx2 \leftarrow idx2 + 1$
12:             **end while**
13:         **end for**
14:         $idx1 \leftarrow idx1 + 1$
15:         **if** $idx1 \leq itl\_inner$ **then**
16:             Resampling
17:         **end if**
18:     **end while**
19:     Update
20: **end for**

---

in a time-step to address the problem of degeneracy [44]. Without resampling, only a small number of particles will have substantial weights for inference.

5. Update: State s_{t+1} is obtained from the resampled particle set $\{s_{t+1}^{(i)}\}_{i=1}^{N_p}$ via weighted average or more complicated functions.

A domain-specific design flow for generating reconfigurable SMC designs is proposed [32]. The design flow minimizes hardware redesign efforts via generic high-level mapping. In other words, application-specific features are specified in a software template and automatically converted to hardware. The SMC computation engine is made up of customizable building blocks and is highly parametrizable to support design optimization.

Fig. 5 shows various steps in the design flow of this approach.

1. Starting with a functional description such as a software code or a mathematical formulation, the users identify and code application-specific features. Generally, only the application-specific features are of interest,

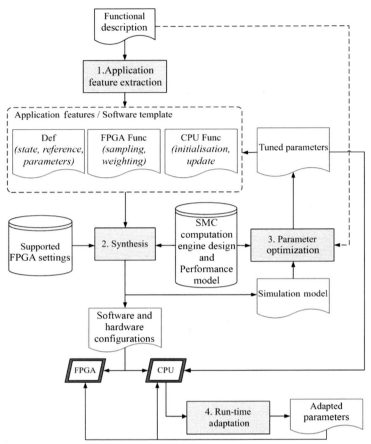

Fig. 5 Design flow (compile time and run time) for SMC applications: Users only customize the application-specific descriptions inside the *dotted box*.

and other features which are common to all SMC applications are handled by the design flow, so the functional description does not necessarily have to be a complete software code.

2. The synthesis step automatically weaves the application–specific features with the computation engine to form a performance model, a simulation model, and a complete configuration for the targeted reconfigurable system. The synthesis tool employed is Maxeler's MaxCompiler. All the application–specific features and the computation engine are described by an extension of Java programming language, which is specialized

for data flow description, such as latency, pipeline, multiplexer, FIFO, and memory. MaxCompiler also supports FPGAs from multiple vendors, such that low-level configurations, such as I/O binding, are performed automatically.

3. The design flow also consists of a parameter optimization step which takes the simulation model and performance model as inputs to produce a set of performance or accuracy optimized parameters. Generally, a simulation model is sufficient for performing optimization, and if a complete software code is provided, it can be used to accelerate the optimization process.

4. The design of SMC computation engine allows further adaptation of design at runtime. The adaptation is based on the solution quality. For example, a better solution quality means that fewer particles could be used for performing SMC and vice versa.

Users create a new SMC design by customizing the application-specific Java descriptions inside the dotted box of Fig. 6. These descriptions correspond to *Def* (Code 1), *DFE Func* (Code 2), and *CPU Func*.

Code 1 illustrates the class, where number representation (floating-point, fixed-point with different bit-width), structs (state, reference), static parameters, and system parameters are defined. Users are allowed to customize number representation to benefit from the flexibility of FPGA and make trade-offs between accuracy and design complexity. State and reference structs determine the I/O interface. Static parameters are defined in this class, while dynamic parameters are provided at runtime. System parameters define device-specific properties such as clock speed and parallelism. Lastly, application parameters define properties that are tied to specific applications.

7.1 DFE Func

Sampling and importance weighting are the most computation intensive functions, and are accelerated by FPGAs. Code 2 gives a simple example on how these two DFE functions are defined. Given current state s_in, reference r_in, and observation m_in (sensor values in this example), an estimation state s_out is computed. Weight w accounts for the probability of an observation from the estimated state. The weight is calculated from the product of scores over the horizon. In this example, the weight is equal to the score as the horizon length is one.

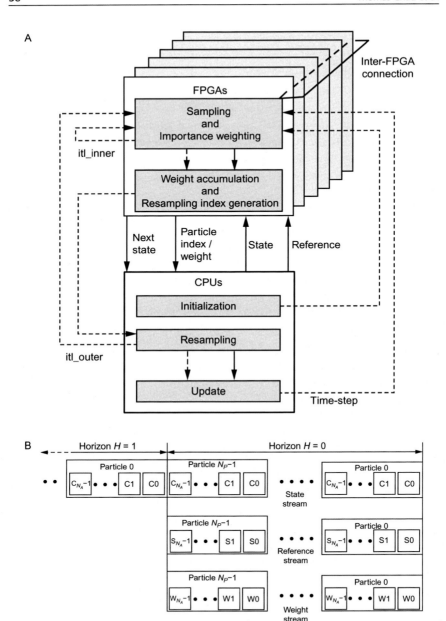

Fig. 6 (A) Design of the SMC computation engine: *Solid lines* represent datapaths, while *dotted lines* represent control paths. (B) Data structure of particles represented by three data streams.

CODE 1 State, Control, and Parameters for the Robot Localization Example

```
public class Def
{
    // Number Representation
    static final DFEType float_t = KernelLib.dfeFloat(8,24);
    static final DFEType fixed_t = KernelLib.dfeFixOffset(26,
-20,SignMode.TWOSCOMPLEMENT);
    // State Struct
    public static final DFEStructType state_t = new
DFEStructType(
        new StructFieldType("x", float_t);
        new StructFieldType("y", float_t);
        new StructFieldType("h", float_t);
    );
    // Reference Struct
    public static final DFEStructType ref_t = new DFEStructType(
        new StructFieldType("d", float_t);
        new StructFieldType("r", float_t);
    );
    // Static Design parameters (Table I)
    public static int NPMin = 5000, NPMax = 25000;
    public static int H = 1, NA = 1;
    // System Parameters
    public static int NC_inner = 1, NC_P = 2;
    public static int Clk_core = 120, Clk_mem = 350;
    public static int FPGA_resampling = 0, Use_DRAM = 0;
    // Application parameters
    public static int NWall = 8;
    public static int NSensor = 20;
}
```

7.2 CPU Func

Initialization and update are functions running on the CPU. They are responsible for obtaining and formatting data and displaying results. *Resampling* is independent of applications so users need not to customize it.

A computation engine has been designed for accelerating SMC applications. To allow customization of the computation engine, the engine

CODE 2 DFE Functions (Sampling and Importance Weighting) for the Robot Localization Example

```
public class Func
{
   public static DFEStruct sampling (DFEStructs_in,
DFEStructc_in){
      DFEStructs_out = state_t.newInstance(this);
      s_out.x = s_in.x + nrand(c_in.d,S*0.5) * cos(s_in.h);
      s_out.y = s_in.y + nrand(c_in.d,S*0.5) * sin(s_in.h);
      s_out.h = s_in.h + nrand(c_in.r,S*0.1);
      return s_out;
   }
   public static DFEVar weighting (DFEStructs_in, DFEVar sensor)
   {
      // Score calculation
      DFEVar score = exp(-1*pow(est(s_in)-sensor,2)/S/0.5);
      // Constraint handling
      bool succeed = est(s_in)>0 ? true : false;
      // Weight accumulation
      DFEVar w = succeed ? score : 0; //weight
      return w;
   }
}
```

and data structure are designed as shown in Fig. 6A and B, respectively. The computation engine employs a heterogeneous structure that consists of multiple FPGAs and CPUs. FPGAs are responsible for sampling, importance weighting, and optionally resampling index generation and are fully pipelined to maximize throughput. To exploit parallelism, particle simulations (sampling and importance weighting) are computed simultaneously by every processing core on each FPGA. Processing cores can be replicated as many times as FPGA resources allow. In situations where the computed results have to be grouped together, data are transferred among FPGAs via an inter-FPGA connection. To maximize the system throughput, remaining noncompute intensive tasks that involve random and nonsequential data accesses are performed on the CPUs. FPGAs and CPUs communicate through high bandwidth connections such as PCI Express or InfiniBand.

From the control paths (dotted lines) of Fig. 6A, we see that there are three loops: (1) inner, (2) outer, and (3) time-step. First, the inner loop iterates *itl_inner* number of times for *sampling* and *importance weighting*, *itl_inner* increases with the iteration count of the outer loop. Second, the outer loop iterates *itl_outer* times to do *resampling*. The resampling process is performed *itl_outer* times to refine the pool of particles. The particle indices are scrambled after this stage, and the indices are transferred to the CPUs to update the particles. Third, the time loop iterates once per time-step to obtain a new control strategy and to update the current state.

Based on this fact, the data structure shown in Fig. 6B is derived. Each particle encapsulates three pieces of information: (1) state, (2) reference, and (3) weight, each being stored as a stream as indicated in the figure. The length of the *state stream* is $N_P \cdot N_A \cdot H$, where H means each control strategy predicts H steps into the future. The *reference* and *weight* streams have information of N_A agents in N_P particles.

The engine design and data structure do not only offer compile-time parameterization but also allow changing the values of *itl_outer*, *itl_inner*, and N_P at runtime. It is because these parameters only affect the length of the particle streams, but not the hardware datapath. The computation engine is fully pipelined and outputs one result per clock cycle.

Fig. 7 shows the design of the FPGA kernel. Blocks that require customization are darkened. The sampling function in Code 2 is mapped to the *Sampling* block which accepts a state and a reference on each clock cycle and calculates the next state on the prediction horizon. After getting a state from the CPU at the beginning (*itl_inner* $= 0$ and $H = 0$), the data will be used by the kernel *itl_inner* $\cdot N_P$ times. An optional *state RAM* enables reuse of state data and improves performance when the value of *itl_inner* is large. An array of LUT-based random number generators [45] is seeded by the CPU to provide random variables, application parameters are stored in registers, and a feedback path stores the state of the previous $N_P \cdot N_A$ cycles.

The *Importance weighting* block computes in three steps. First, *Score calculation* uses the states from the *Next state* block to calculate scores of all the states over the horizon. A feedback loop of length $N_P \cdot N_A$ stores the cost of the previous horizon and accumulates the values. Second, *Constraint handling* uses the states from the *Next state* block to check the constraints. The block raises a fail flag if a constraint is violated. Lastly, *Weight calculation* combines the scores of the states over the horizon.

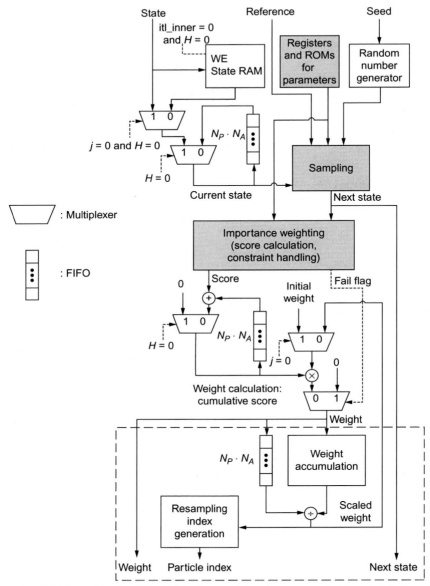

Fig. 7 FPGA kernel design: The blocks that require users' customization are *darkened*. The *dotted box* covers the blocks that are optional on FPGAs.

Part of the resampling process is handled by the *Resampling index generation* and *Weight accumulation* blocks. Weights are accumulated to calculate the cumulative distribution function, and then particles indices are reordered. These two blocks can either be computed on FPGAs or CPUs. All the blocks allow precision customization using fixed-point or floating-point number representation. Users have the flexibility to make trade-off between result accuracy and design complexity.

In summary, this SMC example demonstrates the feasibility of generating highly optimized reconfigurable designs using dataflow hardware, while hiding detailed implementation aspects from the user. A software template makes the computation engine portable and facilitates code reuse, the number of lines of user-written code being decreased for an application.

8. EXAMPLE APPLICATION TWO: SPARSE LINEAR ALGEBRA

Many real life applications need to solve or estimate eigenvalues of sparse systems of linear equations. In this section, we primarily focus on dataflow designs of linear solvers: given a set of linear equations $Ax = b$ with a problem matrix A and known right-hand side vector b, they produce the unknown solution vector x. Typically, the nonzero elements of matrix A and components of vectors are floating point numbers. This problem has high degree of abstraction from the problem domain: the way of generating the matrix and right-hand side vector b, as well as the problem-specific interpretation of the output vector x do not have direct implication to the design of the linear solver. However, the nature of the problem influences the matrix dimensions, its sparsity pattern, and numerical properties. Fig. 8 shows an example of matrix sparsity structure.

For the sparse and unstructured linear problems, iterative methods are known to be well suited, since the only matrix operation they use is matrix–vector multiply. Every computational step of an iterative linear solver has only read access to given matrix and do not require forming new sparse matrix and computing its sparsity pattern. It is especially good property for the dataflow architectures: it avoids dealing with graph data structures, often necessary at the stage of preparing new sparse storage.

For example, one of the most frequently used iterative linear solver methods, conjugate gradient, has the structure shown in Algorithm 2.

Here, similar structured computations are repeated many times until it either reaches the maximum number of iterations or a convergence metric

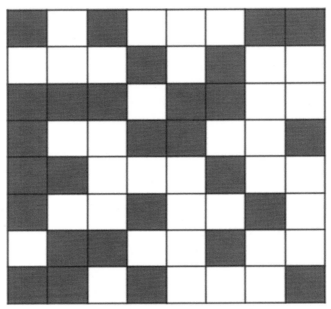

Fig. 8 Example of a matrix sparsity structure. *Black squares* represent nonzero elements.

Algorithm 2 Conjugate gradient method.

1: **function** CONJUGATEGRADIENT(A, x, r)

2:    **for** $i \leq MaxIterations$ **do**

3:        $\vec{A_p} \leftarrow \mathbf{A}\vec{p}$

4:        $\alpha \leftarrow rs_{old}/(\vec{p}^T \vec{A_p})$

5:        $\vec{x} \leftarrow \vec{x} + \alpha\vec{p}$

6:        $\vec{r} \leftarrow \vec{r} - \alpha\vec{A_p}$

7:        $rs_{new} \leftarrow \vec{r}^T \vec{r}$

8:        **if** $rs_{new} < AbsoluteError^2$ **then**

9:            **return** x

10:      **end if**

11:      $\vec{p} \leftarrow \vec{r} + rs_{new}/rs_{old} \times \vec{p}$

12:      $rs_{old} \leftarrow rs_{new}$

13:    **end for**

14: **end function**

Reprinted from G. C. Chow, P. Grigoras, P. Burovskiy, W. Luk, "An efficient sparse conjugate gradient solver using a Beneš permutation network," in: 2014 24th International Conference on Field Programmable Logic and Applications (FPL), IEEE, 2014, pp. 1–7, © 2014 IEEE.

falls below some threshold. In the main loop, the algorithm consists of a sparse matrix–vector multiply (SpMV) (line 3), the number of vector–vector arithmetic operations (lines 5, 6, 11, and 12), and dot products (lines 4 and 7).

Many variations of the conjugate gradient method and other iterative methods have similar structure, with SpMV compute kernel dominating the execution. If the matrix were dense, the matrix–vector multiply compute complexity would be $O(n^2)$, while other compute kernels have complexity $O(n)$. For the sparse matrix, the amount of compute work is proportional to the number n_{nnz} of nonzero entries in the matrix, $n_{nnz} \leq n^2$. Only for the extreme case of exceptionally sparse matrix, the amount of compute work for all other parts of conjugate gradient solver may exceed the complexity of SpMV.

Acceleration of SpMV has often been a challenge. The performance of SpMV computation is significantly limited by the available memory bandwidth due to the difficulty in data reuse and irregular data access patterns. The structure of data access in SpMV kernel is shown in Fig. 9.

Here, every element of the resulting vector is a dot product of a sparse matrix row to a dense vector column. Since the access patterns to the vector column change with every matrix row, the machinery providing efficient data retrieval of vector column data plays the important role to the overall SpMV performance. When it is difficult to retrieve all k vector entries for a

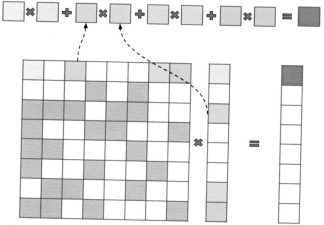

Fig. 9 Multiplication of a first row of sparse matrix to a vector column: every nonzero matrix element (shown in *color*) is multiplied to appropriate vector elements (*different colors* denote respective pairs of matrix and vector elements). Access to vector elements is not linear in its storage and depends on matrix sparsity.

k-wide SpMV arithmetic pipeline, both arithmetic units and memory band-width are under-utilized.

Also, retrieval of *k* vector entries from arbitrary positions in a single execution cycle provides an architectural challenge, which is addressed by maintaining up to *k* copies of vector data on-chip, partitioning vector data into banks, and resolving the bank access conflicts.

Another complication is efficient compressed sparse storage format decoding. Sparse storage schemes store only nonzero matrix elements, along with index data encoding their positions within the matrix. In the common case of a compressed sparse row (CSR) storage format as illustrated in Fig. 10, every nonzero element is represented with 64-bit double precision floating point value and 32-bit column index; additional indexing array encodes the number of nonzero entries in each row, providing their offsets in a contiguous storage of nonzero matrix entries. The number of nonzeros in every matrix row may be different, thus decoder needs to support run length encoding efficiently.

Several dataflow designs proposed recently address the challenges mentioned above. Such challenges are presented below.

8.1 Challenge: Excessive On-chip Memory Usage

A general purpose floating point conjugate gradient solver architecture is proposed in Ref. [46]. This architecture suggests trading the number of vector copies stored on-chip to the amount of data access conflicts by using Beneš permutation network. This network connects the array of arithmetic processing elements (PEs) to the banked vector memory, where each bank stores only a subset of vector *b*. When PEs make their requests for the vector data based on the column indices of matrix nonzeros they process, an access conflict occurs if two (or more) PEs access the same bank. However, if the same subset of vector data is duplicated in few memory banks, the

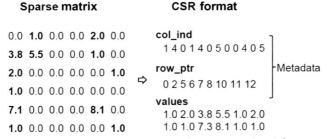

Fig. 10 Sparse matrix is stored in a compressed sparse row (CSR) format.

permutation network may resolve the access conflict at runtime by requesting vector data of same subset from physically different on–chip memories. Thus, the permutation network increases the utilization of arithmetic PEs despite the reduction in the number of data copies on–chip. This type of permutation network is advantageous comparing to the crossbar due to its lower resource usage of $O(N \log N)$ compared to $O(N^2)$ of a crossbar, where N is the number of vector banks.

This architecture is implemented on a single Maxeler Maia acceleration card with an Intel Stratix V FPGA, connected to a Xeon E5-2640 server. The Benes network enables to scale the design up to 128 single precision PEs, each implementing floating point addition, subtraction, and multiplication. Results of performance comparison shows up to 5.9 times speedup vs CPU benchmark (based on Intel MKL Sparse BLAS routines) run on 6–core Xeon X5650 server with HyperThreading (12 threads in total), and up to 9.67 times speedup vs GPU benchmark (based on Cusp framework) on Tesla C2070 card (448 cores).

8.2 Challenge: Avoiding Stalls Due to Data Access Conflicts

Another dataflow architecture proposed in Ref. [47] is a finite element method (FEM) domain-specific design of a conjugate gradient solver. For the FEM problems, sparse matrix A is assembled from a dense block-diagonal matrix M and a sparse incidence matrix A_m of an assembly mapping

$$A = A_m^T M A_m \tag{3}$$

This domain-specific knowledge helps to realize sparse matrix–vector multiply Ax (as part of the conjugate gradient solver) by a chain of compute kernels, implementing block–diagonal dense matrix–vector multiply and (sparse) vector gather/scatter, without explicit formation of a sparse floating point matrix [48]. Fig. 11 shows the data flow diagram of the architecture.

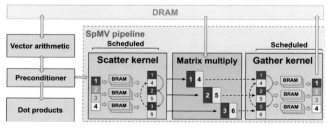

Fig. 11 The structural diagram of the FEM-specific conjugate gradient solver.

In this architecture, SpMV evaluation is realized by a pipeline of three compute kernels: Scatter kernel (block-diagonal), Matrix–vector multiply kernel, and Gather kernel. The multiplicand vector is first scattered to *a larger coefficient* space (an output vector has larger size; some entries of input vector are simply duplicated at the output) to form the input of the matrix multiply kernel. The resulting output vector is reduced to the original size by the Gather kernel, forming the output of SpMV pipeline.

This domain-specific approach encapsulates irregular data access patterns to Gather/Scatter kernels only. Even more, the domain-specific structure of the data access patterns makes it possible to resolve data access conflicts completely via offline scheduling [47].

This dataflow design built for a single Maxeler Maia acceleration card (same as above) is capable of processing FEM meshes with up to one million of so-called *global degrees of freedom* (this is also a dimensions of a sparse matrix A, if it were explicitly formed), which is about 14 times larger than the previous FPGA design of the FEM solver can support [49]. It is also 2.46 times faster than the reference CPU implementation of the similarly structured conjugate gradient solver, a part of the spectral/hp FEM framework Nektar++ [50].

8.3 Challenge: Effective Use of Memory Throughput

A third dataflow design [51] proposes a lossless value compression of the sparse matrix data in addition to compressed sparse storage format. Since the SpMV evaluation is primarily memory bound, the reduction of the matrix dataset is necessary to achieving higher performance with the same DRAM throughput.

The standard sparse storage formats aim to avoid explicit storage of zero entries; however, the array of floating point values of nonzero entries may be further compressed. This design proposes to build a dictionary of only m most frequently occurring nonzero values (as an offline preparation stage) and substitute dictionarized entries to their references into the dictionary. Here, m is a design parameter. Therefore, the original contiguous array of floating point nonzero matrix entries of a compressed sparse storage format (for example, CSR) is replaced with a collection of:
- a bit mask array specifying whether the appropriate matrix element is encoded,
- encoded values in the form of integer indices into a dictionary, and
- the original nonencoded double precision values.

All other indexing arrays of the compressed sparse storage retain their original structure and values, which makes the compression scheme compatible with other sparse storage formats (e.g., CSC, COO, etc.).

Restricting the dictionary size has the following advantages. First, it reduces the on-chip dictionary buffer size, thus making the FPGA decoder potentially very lightweight. Second, when the encoder finds out the most of matrix entries being unique (so that compression leads to increase of dataset rather than reduction), it may set the dictionary size to zero. In the latter case, the compressed storage falls back to the original sparse storage format.

This compression requires a decoder in hardware, which needs to support a run length encoding of matrix values: compressed and noncompressed entries have different bit widths. To address this challenge, Ref. [51] proposes two possibilities: to alternate compressed and noncompressed matrix values, or to split them into the different datasets.

The latter case may be viewed as splitting the original sparse matrix into a sum of two sparse matrices

$$A = A_E + A_U \tag{4}$$

where A_E is completely encoded, and A_U stores the remaining floating point entries. Note that the sparse storage for A_E only contains the indices to the decompression dictionary (along the index of the original sparse storage format), while A_U retains the original sparse storage format. Then, the SpMV evaluation may also be split into stages: first, both matrices are independently multiplied to a vector, then the resulting partial vector results are added together. Keep in mind that splitting the matrices does not increase the amount of compute work.

Ref. [51] implements the latter approach using Maxeler dataflow compiler for both MAX3 and MAX4 DFEs. This implementation deploys a single SpMV circuit and a decoder. The SpMV circuit is used twice: first time to evaluate $A_E x$ (with a decoder), and then to evaluate $A_U x$ (without a decoder). The resource utilization of the double precision decoder is very low: for the MAX3 engine it requires a single BRAM and 16 LUTs and 22 FFs (decoding is nothing but a simple lookup into a static table) for the dictionary of up to 2^{10} entries. For the MAX4 DFE the same dictionary needs 2 BRAMs, 9 LUTs, and 22 FFs. For the larger dictionary with 2^{15} entries the resource utilization grows to 32 BRAMs, 162 LUTs, and 28 FFs on MAX3 and 72 BRAMs, 49 LUTs, and 23 FFs on MAX4.

The compression scheme is evaluated on a number of sparse matrices from University of Florida collection [52]. The maximum achieved compression ratios (normalized with respect to double precision CSR storage) are 17.96, when considering only the nonzero entries. Compared to the entire CSR storage format, the maximum achieved compression ratio is 2.65, with an average between 1.14 and 1.40. Thus, a very low overhead compression technique may give an additional acceleration factor to any SpMV-based architecture.

8.4 Challenge: More Effective On-chip Data Reuse

It is possible to increase the compute intensity of SpMV kernels beyond the bounds of available memory bandwidth by exploiting mathematical structure of the problem. The iterative solvers (e.g., conjugate gradient) perform many iterations with the same sparse matrix, and every iteration requires to fetch sparse matrix data from the external memory again and again. Therefore, the total memory traffic is proportional to the number of iterations necessary for the iterative solver to converge. Recently emerged communication-avoiding solvers [53] propose to merge several iterations of a classical iterative solver with an aim of reducing the number of matrix fetches from the external (thought as slow) memory, at the cost of buffering the matrix data in a faster memory.

A parametrizable dataflow architecture is proposed in Ref. [54] targeting banded sparse linear algebra problems which benefit from building a Krylov subspace

$$\text{span} \left\{ x, Ax, A^x, ..., A^k x \right\} \tag{5}$$

Many linear algebra problems are implicitly or explicitly building the Krylov subspace. Such applications range from iterative linear and eigenvalue solvers and approximations to matrix exponentials to approximate computations of matrix inverse and even Google PageRank citation ranking by the power method [55].

This architecture proposes to deploy up to k SpMV circuits on the same chip and arrange them into a pipeline. The output vector of one SpMV circuit becomes the input to another, which is mathematically equivalent to evaluating

$$Ax^0 = x^1, \ Ax^1 = x^2, ... \tag{6}$$

Due to the data dependencies between consecutive matrix–vector evaluations, SpMV kernels communicate to each other via FIFOs of appropriate capacity. The very first SpMV circuit receives the matrix data from the external (DRAM) memory along with its input vector and passes both to the second SpMV kernel, and so on.

Every SpMV circuit produces its own vector x^i, $i = 1 \ldots k$, forming the basis of the Krylov subspace. Some linear algebra problems need to save all these vectors back to DRAM, but some need only the very last vector $x^{k+1} = A^k x$. A third possibility is that all vectors x^i form the inputs to some other k compute kernels directly on-chip. When only the last element of the Krylov subspace x^{k+1} is needed, the matrix and vector data are transferred from the DRAM only once at the input, and only x^{k+1} is transferred to DRAM at the end of the pipeline.

Typically, the FIFOs required communicating matrix data between kernels are substantially larger then vector FIFOs, thus providing the resource constraint to the maximum depth of the pipeline k and the matrix sizes supported. These matrix buffering overheads have caused by the vector data dependencies between consecutive matrix–vector multiply processes, as presented in Fig. 12.

In Fig. 12 the second SpMV can only start execution when the first SpMV has completed enough elements of its output vector. It also needs to buffer all matrix elements that already contributed to computation of these output vector elements. Note that for the banded matrices (with a relatively small band) the matrix buffer scale as $O(bn)$, where b is matrix band and n is a matrix rank. For a general matrix structure, the on-chip matrix storage could be prohibitively large.

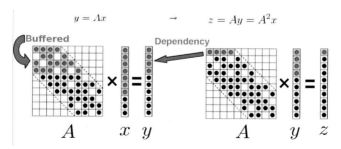

Fig. 12 Data dependency of two consecutive SpMV evaluations: *red dots* represent already processed data. Black dots represent data pending to be processed.

Note all SpMV kernels may perform concurrently as long as they have enough data in their input buffers. The speedup caused by this concurrency (vs independent sequential k evaluations of SpMV) can be expressed as the ratio of runtime of k-independent SpMV evaluations, done sequentially one after another, and the runtime of the Krylov pipeline:

$$S = \frac{k\, T_{\text{SpMV}}}{T_{\text{SpMV}} + (k-1)\, T_{\text{offset}}} = \frac{k n_{nnz}}{n_{nnz} + (k-1)b^2} \tag{7}$$

Since every second SpMV evaluation starts at most b^2 cycles after the first one (to allow the first SpMV circuit computing first b vector entries), the upper bound to the total runtime of the Krylov subspace is the runtime of a single SpMV circuit plus $k-1$ cycle offsets between the activation of the consecutive circuits. Therefore, assuming for simplicity each circuit processes only one matrix element per cycle, the speedup becomes the function of the number of concurrent kernels, the number of nonzero entries in the matrix and the matrix band. Note the requirement for the speedup to be greater than 1 is captured by the inequality $b < \sqrt{n_{nnz}}$.

Fig. 13 presents the graph of the (lower bound to) speedup as a function of the number of SpMV circuits, for a given number of nonzero entries and several matrix bands. Based on graphs of Fig. 13, the Krylov subspace architecture

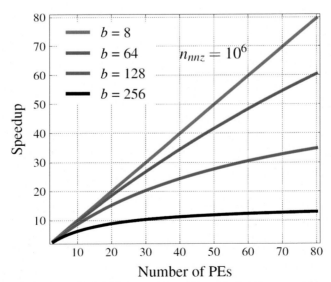

Fig. 13 Speedups show diminishing return as the matrix band squared becomes comparable with the number of nonzero entries.

is expected to scale well with the number of SpMV circuits only for very large matrix problems and moderate matrix bands. For the MAX4 DFE this architecture shows 12.7 (single precision) GFLOPs processing 1.2 GB CSR matrix using 80 internally sequential SpMVs, each capable of processing matrix of band 128. This is 3.3 times faster than the CPU benchmark, running k OpenMP parallel SpMVs (MKL Sparse BLAS) sequentially on a 12-core Xeon server. Note the DFE architecture outperforms the multicore CPU, while DFE's on-board DRAM bandwidth is under-utilized.

9. OTHER APPLICATIONS

DFEs have been applied to a variety of high-performance applications. The following offers a glimpse of how they can be applied to five areas: genomic processing, climate modeling, time series analysis, computer trading strategies, and sparse linear algebra.

9.1 Genomics Processing

Many genomic processing tasks are promising candidates for acceleration using DFEs. The example here is bisulfite sequence alignment, a stage of bioinformatics pipelines for cancer and noninvasive prenatal diagnosis. During sequence alignment, short sequences of DNA letters (reads) are mapped to locations in a known reference genome (Fig. 14). This process currently accounts for over 50% of the analysis time in the pipeline. Alignment of 300 M short reads to the human genome takes roughly 5 h when running on a system with dual 12-core Intel Xeon processors and 100 GB of RAM.

A dataflow system, Ramethy [56], is developed to perform alignment of short reads. Ramethy is implemented on a 1 U Maxeler MPC-X1000

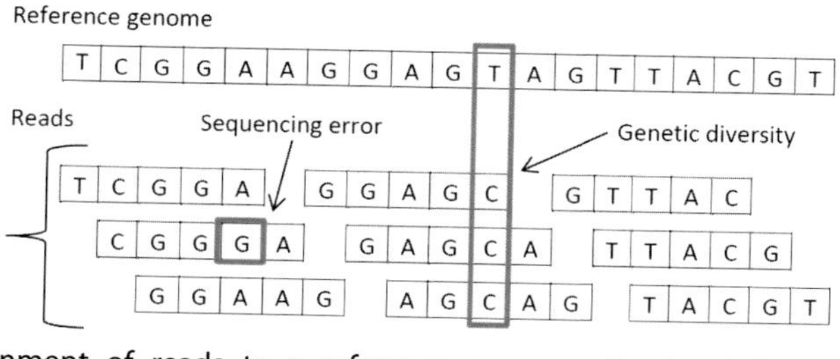

Fig. 14 Alignment of reads to a reference genome. *Reprinted from J. Arram, W. Luk, P. Jiang, "Ramethy: reconfigurable acceleration of bisulfite sequence alignment," in: Proceedings of International Symposium on Field-Programmable Gate Arrays, pp. 250–259, 2015, © 2015 IEEE.*

dataflow node consisting of eight Intel Stratix V FPGAs. A new FM-index structure reduces the number of search steps and computation complexity, and hence improves the pattern matching performance. Results show 15 times speedup compared to dual 12-core Intel Xeon processors, and four times speedup compared to an NVIDIA GTX 580 GPU. In terms of power consumption, Ramethy consumes over an order of magnitude lower energy.

9.2 Climate Modeling

Climate and weather prediction are computationally intensive. Numerical models of weather and climate show significant model error due to limited resolution and complexity, necessitating a need for even more resource-intensive models. It is a challenge to resolve important convective cloud systems in global models, and at the same time meeting performance and power requirements for running such models.

A dataflow architecture for modeling atmospheric variables with chaotic behavior has been developed [57]. Reduced precision optimization is applied to a two-scale Lorenz'96 simulation. Hellinger distance is used to compare statistical behavior between reduced-precision implementations to make trade-offs between precision and throughput. Results show that a Maxeler MAX3A DFE consisting of a Xilinx Virtex-6 FPGA can be 13 times faster and 23 times more power efficient than a 6-core Intel Xeon X5650 processor.

9.3 Time Series Analysis

Ordinal analysis is a statistical method for analyzing the complexity of time series. This method has been used in characterizing dynamic changes in time series, with various applications such as financial risk modeling and biomedical signal processing. It is computationally demanding particularly for high query orders and large time series data. A dataflow system is designed to accelerate ordinal pattern encoding [58]. The challenge is the operations involve sequence sorting that do not have efficient hardware implementations. A two-level ordinal pattern encoding scheme is proposed to avoid sequence sorting and reduce data transmission. The system is implemented on a Maxeler MAX3 DFE which shows substantial acceleration compared to software solutions.

9.4 Computer Trading Strategies

In developing computer trading strategies, genetic programming has been applied to enable the recognition of complex market patterns and behaviors. Genetic programming involves repeatedly generating a set of programs, evaluating them on a large data set and selecting the best performing ones. The evaluation step computes a fitness metric for each program, based on which the best performing programs can be selected for the next iteration. The potentially complex programs and large data sets on which they need to be evaluated make fitness evaluation one of the most computationally expensive components in genetic programming and make genetic programming an unfeasible technique in the context of high-frequency markets.

A dataflow system [59] is proposed to accelerate the fitness evaluation of a genetic program, enabling identification of complex data patterns such as those within foreign exchange data which could lead to more advanced trading strategies. The system uses a pipelined architecture for evaluating the fitness function of complete expression trees representing trading rules (Fig. 15) and supports both fixed point and single-precision floating-point arithmetic. Demonstration with sets of synthetic and real market data shows a speedup of up to 22 times when compared to an optimized 12-core CPU implementation.

9.5 Real-Time Proximity Query

Proximity query (PQ) is a process to calculate the relative placement of objects modeled by vertices with complex morphology. It is a critical task for many applications such as robot motion planning and haptic rendering, but it is often too computationally demanding for real-time applications, particularly those involving human-robot collaborative control. In Ref. [60]

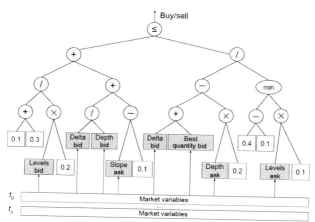

Fig. 15 Expression tree for a trading rule. *Reprinted from A.I. Funie, P. Grigoras, P. Burovskiy, W. Luk, M. Salmon, "Reconfigurable acceleration of fitness evaluation in trading strategies," in: Proceedings of International Conference on Application-Specific Systems, Architectures and Processors, 2015, © 2015 IEEE.*

a PQ formulation is implemented on DFEs. To leverage the advantages of FPGAs, function transformation eliminates iterative trigonometric functions such that the algorithm can be fully pipelined. Datapath parallelism is increased by adopting a reduced precision data format which consumes fewer logic resources than high precision. To maintain the accuracy of results, potential incorrect outputs are recomputed in high precision. Run-time reconfigurability of FPGA is exploited to optimize precision dynamically. Experimental results show that the optimized PQ implementation on a dataflow platform with four DFEs is substantially faster than an optimized CPU, GPU, and a double precision FPGA implementation.

9.6 Stencil Computation

Stencil computation is widely used in diverse areas such as heat diffusion, electromagnetic, and fluid dynamics. By sweeping over a spatial grid, the stencil kernel performs nearest neighboring computation in multiple dimensions. Stencil computation is computationally intensive, especially when the number of dimensions increases, memory access becomes sparser and the achievable throughput is being limited. In Ref. [61] DFEs are used as field–programmable accelerators for stencil computation, with memory architectures for a single stencil operator and for parallel stencil operators (Fig. 16). A partitioning algorithm extracts valid and optimized partitions

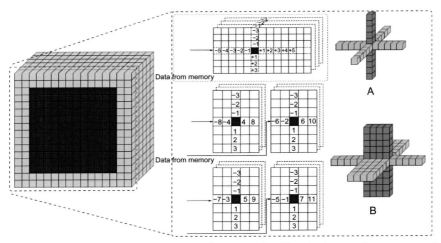

Fig. 16 Stencil computation. Memory architectures for (A) a single stencil operator and (B) parallel stencil operators. *Reprinted from X. Niu, Q. Jin, W. Luk, Q. Liu, O. Pell, "Exploiting run-time reconfiguration in stencil computation," in: Proceedings of International Conference on Field Programmable Logic and Applications, pp. 173–180, 2012, © 2012 IEEE.*

to exploit run-time potential and to achieve optimal system acceleration. An analytical model is used to enable design space exploration for optimizing memory architectures, precision, and design scale. The approach is applied to a reverse time migration application, showing improved speed and resource usage.

9.7 Genetic Algorithms

Genetic algorithms are a class of numerical and combinational optimizers which are especially useful for solving complex nonlinear and nonconvex problems. Effort to apply genetic algorithms to large scale and latency-sensitive problems has been made in Ref. [62]. Pipelined genetic propagation is proposed to map the algorithm to DFEs efficiently and address the challenge in large shared population memory and intrinsic loop-carried dependency. Such DFE-oriented approach is intrinsically distributed and pipelined. The genetic algorithm solver is represented as a graph of loosely coupled genetic operators, which allows the solution to be scaled to the available resources and to dynamically change topology at run-time to explore different solution strategies. Experiments show the DFE is effective in accelerating different applications that apply genetic algorithms.

10. SUMMARY

The year 2015 marks the 40th anniversary of introduction of dataflow processors. Much progress on development of dataflow systems has been made in the last 40 years, with DFEs beginning to be deployed for research and for commercial applications. In particular, dataflow systems have recently been shown to provide significant benefits in performance and in energy efficiency for a variety of high-performance designs that address global challenges, such as genomic analysis and climate modeling. These give us confidence that advances in dataflow systems would be even more exciting for the next 40 years.

ACKNOWLEDGMENT

The support of EPSRC Grant EP/I012036/1, Imperial College London Pathway to Impact award P52806, and the European Union Horizon 2020 Programme under Grant Agreement Number 671653 is gratefully acknowledged. We thank Dr. Oskar Mencer for comments that greatly improved the manuscript.

REFERENCES

[1] A.H. Veen, Dataflow machine architecture, ACM Comput. Surv. 18 (4) (1986) 365–396.
[2] B. Lee, A.R. Hurson, Dataflow architectures and multithreading, Computer 27 (8) (1994) 27–39.
[3] A.R. Hurson, K.M. Kavi, Dataflow computers: their history and future, Wiley Encyclopedia of Computer Science and Engineering, John Wiley & Sons, Inc., Hoboken, NJ, 2008.
[4] https://www.microsoft.com/en-us/research/project/project-catapult/. Retrieved 9 Feb 2017.
[5] http://press.xilinx.com/2016-10-17-Baidu-Adopts-Xilinx-to-Accelerate-Machine-Learning-Applications-in-the-Data-Center. Retrieved 9 Feb 2017.
[6] https://aws.amazon.com/ec2/instance-types/f1/. Retrieved 9 Feb 2017.
[7] G.M. Amdahl, Validity of the single processor approach to achieving large-scale computing capabilities, AFIPS Conference Proceedings, vol. 30, 1967, pp. 483–485.
[8] D.L. Slotnick, Unconventional systems, AFIPS Conference Proceedings, vol. 30, 1967, pp. 477–481.
[9] M.J. Flynn, P. Hung, Microprocessor design issues: thoughts on the road ahead, IEEE Micro 25 (3) (2005) 16–31.
[10] J.B. Dennis, D.P. Misunas, in: A preliminary architecture for a basic data-flow processor, Proceedings of Annual Symposium on Computer Architecture, 1975, pp. 126–132.
[11] Arvind, D.E. Culler, Dataflow architectures, Ann. Rev. Comput. Sci. 1 (1986) 225–253.
[12] P.C. Treleaven, D.R. Brownbridge, R.P. Hopkins, Data-driven and demand-driven computer architecute, ACM Comput. Surv. 14 (1) (1982) 93–143.
[13] J.R. Gurd, C.C. Kirkham, I. Watson, The manchester prototype dataflow computer, Commun. ACM 28 (1) (1985) 34–52.
[14] Arvind, R.S. Nikhil, Executing a program on the MIT tagged-token dataflow architecture, IEEE Trans. Comput. 39 (3) (1990) 300–318.
[15] Y.N. Patt, W.W. Hwu, M. Shebanow, HPS, a new microarchitecture: rationale and introduction, Proceedings of Annual Workshop on Microprogramming, 1985.
[16] D. Burger, S.W. Keckler, K.S. McKinley, M. Dahlin, L.K. John, C. Lin, et al., Scaling to the end of silicon with EDGE architectures, Computer 37 (7) (2004) 44–55.
[17] http://www-304.ibm.com/webapp/set2/sas/f/capi/home.html. Retrieved 9 Feb 2017.
[18] J.R. Gurd, The manchester dataflow machine, Comput. Phys. Commun. 37 (1) (1985) 49–62.
[19] https://www.altera.com/products/design-software/embedded-software-developers/opencl/documentation.html. Retrieved 9 Feb 2017.
[20] http://www.maxeler.com. Retrieved 9 Feb 2017.
[21] O. Pell, V. Averbukh, Maximum performance computing with dataflow engines, Comput. Sci. Eng. 14 (2012) 98–103.
[22] https://www.altera.com/products/design-software/fpga-design/quartus-prime/overview.html. Retrieved 9 Feb 2017.
[23] http://www.xilinx.com/products/design-tools/vivado.html. Retrieved 9 Feb 2017.
[24] X. Niu, T.C.P. Chau, Q. Jin, W. Luk, Q. Liu, O. Pell, Automating elimination of idle functions by runtime reconfiguration, ACM Trans. Reconfigurable Technol. Syst. 8 (3) (2015).
[25] X. Niu, Q. Jin, W. Luk, S. Weston, A self-aware tuning and self-aware evaluation method for finite-difference applications, ACM Trans. Reconfigurable Technol. Syst. 7 (2) (2014).
[26] J.G.F. Coutinho, O. Pell, E. O'Neill, P. Sanders, J. McGlone, P. Grigoras, et al., in: HARNESS project: managing heterogeneous computing resources for a cloud platform, Proceedings of International Symposium on Applied Reconfigurable Computing, 2014, pp. 324–329.

[27] P. Grigoras, X. Niu, J.G.F. Coutinho, W. Luk, J. Bower, O. Pell, in: Aspect driven compilation for dataflow designs, Proceedings of International Conference on Applications Specific System Architectures and Processors, 2013, pp. 18–25.

[28] J.M.P. Cardoso, T. Carvalho, J.G.F. Coutinho, W. Luk, R. Nobre, P. Diniz, et al., in: LARA: an aspect-oriented programming language for embedded systems, Proceedings of International Conference on Aspect-Oriented Software Development, 2012, pp. 179–190.

[29] P. Grigoras, M. Tottenham, X. Niu, J.G.F. Coutinho, W. Luk, in: Elastic management of reconfigurable accelerators, Proceedings of International Symposium on Parallel and Distributed Processing with Applications, 2014, pp. 174–181.

[30] E. O'Neill, J. McGlone, J.G.F. Coutinho, A. Doole, G. Ragusal, O. Pell, et al., in: Cross resource optimisation of database functionality across heterogeneous processors, Proceedings of International Symposium on Parallel and Distributed Processing with Applications, 2014, pp. 150–157.

[31] S.N. Hmid, J.G.F. Coutinho, W. Luk, in: A transfer-aware runtime system for heterogeneous asynchronous parallel execution, Proceedings of International Symposium on Highly-Efficient Accelerators and Reconfigurable Technologies, 2015.

[32] T.C.P. Chau, M. Kurek, J.S. Targett, J. Humphrey, G. Skouroupathis, A. Eele, et al., in: SMCGen: generating reconfigurable design for sequential Monte Carlo applications, Proceedings of International Symposium on Field-Programmable Custom Computing Machines, 2014, pp. 141–148.

[33] A. Doucet, N. de Freitas, N. Gordon, Sequential Monte Carlo Methods in Practice, Springer-Verlag, New York, 2001.

[34] M. Happe, E. Lubbers, M. Platzner, A self-adaptive heterogeneous multi-core architecture for embedded real-time video object tracking, J. Real Time Image Process 8 (1) (2013) 95–110. Springer-Verlag New York, Inc., Secaucus, NJ.

[35] M. Montemerlo, S. Thrun, W. Whittaker, in: Conditional particle filters for simultaneous mobile robot localization and people-tracking, Proceedings of International Conference Robotics and Automation, 2002, pp. 695–701.

[36] T.C.P. Chau, X. Niu, A. Eele, J.M. Maciejowski, P.Y.K. Cheung, W. Luk, Mapping adaptive particle filters to heterogeneous reconfigurable systems, ACM Trans. Reconfig. Technol. Syst. 7 (4) (2014).

[37] J. Vermaak, C. Andrieu, A. Doucet, S.J. Godsill, Particle methods for Bayesian modeling and enhancement of speech signals, IEEE Trans. Speech Audio Process. 10 (3) (2002) 173–185.

[38] T.C.P. Chau, J.S. Targett, M. Wijeyasinghe, W. Luk, P.Y.K. Cheung, B. Cope, et al., Accelerating sequential Monte Carlo method for real-time air traffic management, Comput. Archit. News 41 (5) (2013) 35–40.

[39] A. Eele, J.M. Maciejowski, T.C.P. Chau, W. Luk, in: Parallelisation of sequential Monte Carlo for real-time control in air traffic management, Proceedings of International Conference Decision and Control, 2013.

[40] A. Eele, J.M. Maciejowski, T.C.P. Chau, W. Luk, in: Control of aircraft in the terminal manoeuvring area using parallelised sequential Monte Carlo, Proceedings of AIAA Conference on Guidance, Navigation, and Control, 2013.

[41] N. Kantas, J.M. Maciejowski, A. Lecchini-Visintini, Sequential Monte Carlo for model predictive control, in: Nonlinear Model Predictive Control, vol. 384, Lecture Notes in Control and Information Sciences, 2009, pp. 263–273.

[42] D. Creal, A survey of sequential Monte Carlo methods for economics and finance, Econ. Rev. 31 (3) (2012) 245–296.

[43] N.J. Gordon, D.J. Salmond, A.F.M. Smith, Novel approach to nonlinear/non-Gaussian Bayesian state estimation, IEE Proc. F Radar Signal Process. 140 (2) (1993) 107–113.

[44] G. Kitagawa, Monte Carlo filter and smoother for non-Gaussian nonlinear state space models, J. Comput. Graph. Stat. 5 (1) (1996) 1–25.

[45] D.B. Thomas, W. Luk, High quality uniform random number generation using LUT optimised state-transition matrices, J. Signal Process. Syst. 47 (1) (2007) 77–92.

[46] G.C. Chow, P. Grigoras, P. Burovskiy, W. Luk, in: An efficient sparse conjugate gradient solver using a Beneš permutation network, 2014 24th International Conference on Field Programmable Logic and Applications (FPL), IEEE, Munich, Germany, 2014, pp. 1–7.

[47] P. Burovskiy, P. Grigoras, S. Sherwin, W. Luk, Efficient assembly for high order unstructured FEM meshes, 2015 25th International Conference on Field Programmable Logic and Applications (FPL), IEEE, London, UK, 2015, pp. 36–41.

[48] G. Karniadakis, S. Sherwin, Spectral/hp Element Methods for Computational Fluid Dynamics, Oxford University Press, Oxford, UK, 2013.

[49] J. Hu, S.F. Quigley, A. Chan, in: An element-by-element preconditioned conjugate gradient solver of 3d tetrahedral finite elements on an fpga coprocessor, International Conference on Field Programmable Logic and Applications, 2008. FPL 2008, IEEE, Heidelberg, Germany, 2008, pp. 575–578.

[50] C. Cantwell, et al., Nektar++: an open-source spectral/hp element framework, Comput. Phys. Commun. 192 (2015) 205–219.

[51] P. Grigoras, P. Burovskiy, E. Hung, W. Luk, in: Accelerating SpMV on FPGAs by compressing nonzero values, 2015 IEEE 23rd Annual International Symposium on Field-Programmable Custom Computing Machines (FCCM), IEEE, Vancouver, BC, Canada, 2015, pp. 64–67.

[52] T.A. Davis, Y. Hu, The University of Florida sparse matrix collection, ACM Trans. Math. Softw. 38 (1) (2011) 1.

[53] A. Rafique, G. Constantinides, N. Kapre, Communication optimization of iterative sparse matrix-vector multiply on GPUs and FPGAs, IEEE Trans. Paral. Distrib. Syst. 26 (1) (2015) 24–34.

[54] P. Burovskiy, S. Girdlestone, C. Davies, S. Sherwin, W. Luk, in: Dataflow acceleration of Krylov subspace sparse banded problems, Proceedings of International Conference on Field Programmable Logic and Applications, 2015, pp. 1–6.

[55] L. Page, S. Brin, R. Motwani, T. Winograd, "The PageRank Citation Ranking: Bringing Order to the Web", Tech. rep, Stanford Digital Library Technologies Project1998.

[56] J. Arram, W. Luk, P. Jiang, in: Ramethy: reconfigurable acceleration of bisulfite sequence alignment, Proceedings of International Symposium on Field-Programmable Gate Arrays, 2015, pp. 250–259.

[57] F.P. Russell, P.D. Duben, X. Niu, W. Luk, T.N. Palmer, in: Architectures and precision analysis for modelling atmospheric variables with chaotic behavior, Proceedings of International Symposium on Field-Programmable Custom Computing Machines, 2015, pp. 171–187.

[58] C. Guo, W. Luk, S. Weston, in: Pipelined reconfigurable accelerator for ordinal pattern encoding, Proceedings of International Conference on Architectures and Processors, 2014, pp. 194–201.

[59] A.I. Funie, P. Grigoras, P. Burovskiy, W. Luk, M. Salmon, in: Reconfigurable acceleration of fitness evaluation in trading strategies, Proceedings of International Conference on Application-Specific Systems, Architectures and Processors, 2015.

[60] T.C.P. Chau, K.W. Kwok, G.C.T. Chow, K.H. Tsoi, Z. Tse, P.Y.K. Cheung, et al., in: Acceleration of real-time proximity query for dynamic constraints, Proceedings of International Conference on Field-Programmable Technology, 2013.

[61] X. Niu, Q. Jin, W. Luk, Q. Liu, O. Pell, in: Exploiting run-time reconfiguration in stencil computation, Proceedings of International Conference on Field Programmable Logic and Applications, 2012, pp. 173–180.

[62] L. Guo, C. Guo, D.B. Thomas, W. Luk, in: Pipelined genetic propagation, Proceedings of International Symposium on Field-Programmable Custom Computing Machines, 2015, pp. 103–110.

ABOUT THE AUTHORS

Thomas Chau is an Electronic Design Engineer at Intel Corporation. His work focuses on optimizing real-time applications with novel hardware such as FPGAs. He obtained a PhD in Computing at Imperial College London in 2014. Before that, he received BEng and MPhil degrees in Computer Science and Engineering from the Chinese University of Hong Kong in 2008 and 2010, respectively.

Pavel Burovskiy is a Dataflow Software Engineer at Maxeler Technologies with focus on HPC applications with unstructured data. Previously, postdoc at Imperial College London (Department of Computing, Department of Aeronautics) and University of Exeter. PhD in Mathematics.

Wayne Luk is a Professor of Computer Engineering at Imperial College London. He leads the Custom Computing Research Group and the EPSRC Centre for Doctoral Training in High Performance Embedded and Distributed Systems. His current research interests include theory and practice of customizing hardware and software for specific application domains such as machine learning, genomics and climate modeling, and high-level compilation techniques and tools for high-performance computers and embedded systems. He is a fellow of the Royal Academy of Engineering, the IEEE, and the BCS. He is a recipient of the Research Excellence Award from Imperial College London, and many awards for his publications from various international conferences.

Michael J. Flynn, Professor of Electrical Engineering at Stanford University, is best known for the [SIMD, SISD, MISD, MIMD] classification and the first detailed discussion of super scalar design. He was founder and senior consultant to Palyn Associates, a leading computer design company; founder and Vice President of American Supercomputers; and a partner at Paragon Partners, a venture capital partnership. Prof. Flynn received the IEEE/ACM Eckert-Mauchley and Harry Goode Memorial Awards in 1992 and 1995, respectively.

CHAPTER THREE

Adaptation and Evaluation of the Simplex Algorithm for a Data-Flow Architecture

Uroš Čibej, Jurij Mihelič
Faculty of Computer and Information Science, University of Ljubljana, Ljubljana, Slovenia

Contents

Advances in Computers, Volume 106
ISSN 0065-2458
http://dx.doi.org/10.1016/bs.adcom.2017.04.003

63

Abstract

The main goal of this chapter is to present a novel adaptation of the classical simplex algorithm for a data-flow architecture. Due to the recent reemergence of the data-flow paradigm, most of the algorithms need to be reengineered in order to use the features of the new platform. By exploring various possibilities of implementations and by extensive empirical testing we manage to show the suitability of the data-flow paradigm for the simplex algorithm, as well as pinpoint the strengths and some of the weaknesses of the used architecture.

ABBREVIATIONS

BRAM block random access memory
CPU central processing unit
DFE data-flow engine
DSP digital signal processing slices
FF flip flop
FMem fast memory
FPGA field-programmable gate array
LMem large memory
LUT lookup table
LP linear programming
PCI peripheral component interconnect
PCIe PCI Express

1. INTRODUCTION

Reconfigurable hardware is emerging as a practical platform for the introduction of new computational paradigms. This is especially true for high-performance reconfigurable hardware, where the flexibility can be exploited to better suit the needs of the underlying problem. One of the computational models which was brought back to life by reconfigurable hardware is the data-flow paradigm [1]. The data-flow architecture is one of the alternatives to the now omni-present von Neumann architecture, which has not established itself as a viable alternative in the past, mainly due to technological constraints which made it less practical. The recent technological advances, driven mainly by Maxeler [2], made the data-flow paradigm not only competitive with the control-flow-based processors, but overtaking them in many different aspects [3]. The most important aspect is the possibility to manipulate huge amounts of data while at the same time using significantly less energy than comparable control-flow-based solutions

[4,5]. In the era of Big Data and growing concerns about the energy efficiency, this aspect makes data-flow a very appealing solution.

But the speedups which can be achieved with this approach are unfortunately not straightforward. To be able to fully exploit the advantages of the data-flow architecture, the algorithms have to be carefully reengineered, since most algorithms were tailored specifically for the control-flow architecture and are thus often not directly implementable as a data-flow. Many algorithms have already been implemented for this architecture, from sorting algorithms [6,7], fast Fourier transform [8] to various physics simulations, e.g., weather models [9]. It has been most successful in classical domains of high-performance computing, i.e., various numerical applications. Our global goal is to experiment with other domains, thus finding candidates for a successful utilization of this technology.

The focus of this chapter is linear programming (LP), which is one of the most widely used paradigm for modeling linear optimization problems. One of the main reasons for such popularity of LP is the existence of an extremely fast algorithm for solving (optimally) very large instances of problems. The algorithm is known as the simplex algorithm. Ranking algorithms is definitely a difficult and very subjective task, but almost in any ranking the simplex algorithm is deemed as one of the most influential algorithms in history [10]. Our goal is to demonstrate the suitability of this algorithm for the data-flow architecture, by significantly improving the running times as compared to the control-flow implementation.

The majority of LP solvers implement the revised simplex method, since the revised simplex method is more suitable for sparse linear programs. In this chapter we focus on dense linear programs, which appear in several practical applications [11,12]. For this type of linear programs the primal simplex algorithm is equally efficient [13] (and sometimes even more) as the revised simplex. But the simplicity of the primal simplex method makes is much more amenable to parallelization and during our implementations it turned out to be more suitable for the implementation on a data-flow architecture.

The rest of the chapter is structured as follows. The next section gives a short introduction to the data-flow architecture. Section 3 describes the LP problem and an overview of the simplex algorithm. Section 4 describes two implementations of this algorithm on the data-flow architecture. The first implementation retains all the data in the main computer memory, while the second implementation takes advantage of the on-chip memory. Section 5 gives an empirical evaluation of the presented algorithms, comparing them to an optimized central processing unit (CPU) implementation as well as a comparison with a state-of-the-art LP solver, demonstrating the superiority of the data-flow implementations. Finally, Section 6 concludes the chapter, giving a discussion of the presented results and the directions for future research in further improving the simplex algorithm.

2. THE MAXELER ARCHITECTURE

Due to the predominance of the von Neumann (control-flow) architecture in today's computer systems, the data-flow computers are usually seen as a rather obscure computational model. Even though the data-flow paradigm has been around for as much time as the control flow, it has never been a viable competitor for practical applications. However, in recent years, the von Neumann architecture is experiencing a tough technological barrier. This is pushing the researchers and hardware producers to experiment with a wide variety of alternative platforms.

In this chapter we experiment with a specific solution, namely the Maxeler platform, since it has shown very encouraging results on many practical applications. The Maxeler approach to the data-flow is pragmatic, i.e., it is not trying to build a general purpose data-flow computer, but rather use it as an accelerator for control-flow processors. A similar approach has been taken by massively parallel processors such as general purpose graphics processing units and has been very successful. Such hybrid approach can exploit the advantages of both paradigms, making it more successful on a wider variety of applications.

The enabling technology for the Maxeler platform is the advent of field-programmable gate arrays (FPGAs), which is being used as the target for the data-flow layout. Since the data-flow is physically laid out in space (on the FPGA chip), this type of computing is also referred to as spatial computation.

In what follows we give an overview of the platform, omitting many details, but giving enough information for the reader to understand the concepts that will be used and discussed in the remainder of this chapter. We are particularly interested in the abstract overview of the entire system, focusing on the details that are most relevant for developers of high-performance applications. A more thorough introduction to the Maxeler architecture can be found in Ref. [14].

2.1 Structural Overview

As mentioned earlier, the platform encompasses both the control-flow as well as the data-flow paradigm. These two are encompassed in separate computational units:

- A control-flow unit is responsible for the setup and supervision of the data-flow engine. Depending on the problem it does some (pre/post) processing of the data.
- A data-flow engine (DFE)—the unit where the data-flow part of the computation is executed. This is basically the FPGA with some additional memory and logic. The data-flow on this FPGA is obtained from a set of Maxeler tools, which are responsible for compilation as well as the orchestration of the execution of a particular data-flow.

The two units communicate through input and output ports, all the communication being initiated by the CPU. In a nutshell, the execution of a program on this platform works as follows:

1. the CPU creates a set of streams (in its main memory),
2. pushes them through to the DFE, and
3. receives the transformed data back to its main memory.

These three steps can be executed several times, depending on the application at hand. However, the data transfer to/from the DFE is time consuming. In order to overcome this bottleneck, the DFE has a lot of local memory. Some intermediate results can be saved only on the DFE, thus reducing the transfer times between the two computational units. The memory on the DFE is of two types. The first type of memory is the large memory (LMem), which is a typical random access memory. The second type of memory is the fast memory (FMem), which is implemented on the FPGA. For the programmer these two types of memory differ in two important aspects. The first difference is the speed, LMem having a bandwidth in the order of GB/s, whereas the FMem having the bandwidth in the order of TB/s. The second difference is the size, FMem being very limited, typically a few megabytes, compared to several gigabytes for LMem. These differences guide the development of applications, searching for a compromise between the speed and the capacity of the available memory.

2.2 Programmers View

In order for a programmer to devise a complete program, three components need to be written.

CPU code: typically written in the C programming language, the CPU code controls the execution and uses the DFE as a processing unit by calling suitable functions exposed by the Maxeler compiler.

Set of kernels: each kernel implements a certain functionality and is roughly an equivalent of a function abstraction. It has a set of input streams and a set of output streams attached.

Manager: The manager is the component that connects the data streams from the CPU to the recipient kernels and vice versa. It establishes connections between the kernels and the LMem as well as interconnects the kernels. The manager also constructs the interfaces with which the CPU code interacts with the DFE.

The manager and the kernels are written in a domain-specific language called MaxJ. This language is a superset of the Java programming language, with a few extensions which are more suitable for an easier creation of the data-flow programs.

The compiler transforms the description of the kernels into a data-flow graph and this graph is physically laid out on the FPGA chip by the backend. The backend is typically very computationally intensive, since there are many structural constraints to be taken into account.

A schematic view of this architecture can be observed in Fig. 1.

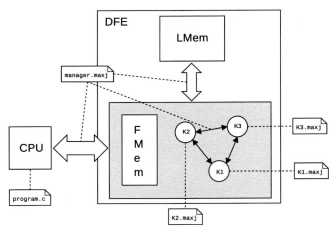

Fig. 1 A schematic overview of the components of the data-flow system. The *gray box* represents the FPGA chip. The programmers view of the architecture is shown as the files that need to be implemented for every program, the .maxj file are written in the MaxJ language, whereas the control flow is typically written in C or C++, but also other languages are being supported.

3. LINEAR PROGRAMMING

In this section we give a detailed description of the LP problem and a few basic concepts, which impact the algorithmic complexity. We continue by giving an overview of the simplex algorithm and its implementation on the data-flow architecture.

3.1 Problem Definition

As mentioned earlier, LP is a basic mathematical tool in a wide variety of disciplines, from airline industry [15], finance [16] to logistics [17], supply-chain management [18], and many others.

The basic concepts of a linear program are variables, representing different quantities and linear constraint, where these variables are combined together and bounded with an (in)equality. A linear combination of these variables defines an objective function which is to be minimized or maximized. Let us demonstrate the introduced concepts with an example.

Example 1. As an example, let us look at a simple linear program involving two variables (x and y), subject to (s.t.) four explicit constraints and two implicit constraints $x, y \geq 0$. The objective is to maximize the function in the first line below.

$$
\begin{aligned}
\text{Max} \quad & x + 2y \\
\text{s.t.} \quad & x - 5y \leq 0 \\
& -x + 5y \leq 15 \\
& x - y \leq 2 \\
& x + y \leq 8 \\
& x, y \geq 0.
\end{aligned}
$$

The constraints can be visualized as a two-dimensional convex polygon as shown in Fig. 2. All feasible solutions are contained in the area marked as F. The goal function to be maximized is visualized as the dotted lines, each dotted line represents points with an equal value of the goal function. The arrow represents the optimal point in the region F and the direction in which the goal function grows even further.

3.2 The Standard Form

The linear program given in Example 1 is in the, so-called, standard form. This means that all the constraints are in the form of $\leq$, all the variables are nonnegative, and the objective function is to be maximized. Even though

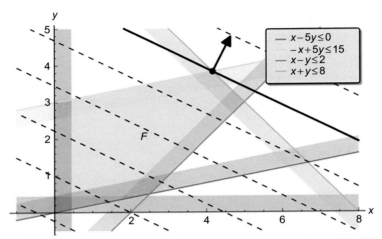

Fig. 2 A two-dimensional linear program, with the feasible region denoted by *F*. The *dot* indicates the optimal solution, and the *arrow* indicates the direction of growth of the objective function.

this might seem as a reduced set of linear programs, with a few simple transformations any linear program can be transformed into the standard form. Here is a list of transformations.

- A minimization can be transformed into a maximization problem by simply multiplying the objective function by -1.
- The inequalities of the form $\geq$ can be transformed to the $\leq$ form by multiplying both sides of the inequality by -1.
- The constraints in the form of equalities are transformed into two inequalities, one $\geq$ and one $\leq$.
- Variables with lower bounds $x \geq l$, where $l < 0$, are substituted with $x = w + l$ and the new variable w has the correct nonnegativity constraint.
- Similarly, variables with an upper bound $x \leq u$ are substituted with $x = u - w$.
- Free variables, i.e., unbounded variables are substituted with $x = u - v$, where u and v are new nonnegative variables.

Example 2. As an example of these transformation we transform the following linear program into the standard form.

$$
\begin{array}{rlrl}
\text{Min} & 3x_1 & + & 8x_2 \\
\text{s.t.} & -x_1 & + & x_2 & \geq & 10 \\
& -x_1 & + & 2x_3 & = & 10 \\
& & & x_{1,2} & \geq & 0 \\
& & & x_3 & \leq & 9.
\end{array}
$$

With the above described transformations, we get the following standard form:

$$
\begin{aligned}
\text{Max} \quad -3x_1 \; - \; & 8x_2 \; + \; (-36) \\
\text{s.t.} \qquad x_1 \; - \; & x_2 \; \leq \; -10 \\
-x_1 \; - \; & 2x_4 \; \leq \; -8 \\
x_1 \; + \; & 2x_4 \; \leq \; 8 \\
& x_{1,2,4} \; \geq \; 0.
\end{aligned}
$$

We introduced a new variable x_4 due to the substitution $x_3 = 9 - x_4$. The constant (-36) introduced into the objective function can be ignored, since the optimal solution is the same. The equality in the original program is transformed into two inequalities.

In the remainder of this chapter we will assume the program to be in the standard form. When discussing the time complexities, especially in Section 5, the transformations into the standard form will not be taken into account. But since the time taken for the transformation is an order of magnitude smaller than the time to find the optimal solution, this is a reasonable assumption to make.

3.3 Duality

Another important concept (besides the standard form) in LP is the duality of a linear program. In general optimization, the dual optimization problem is a problem that provides an upper bound to the original maximization problem (which we will call primal). But in LP the optimal solution to the dual (minimization) problem provides the exact solution to the primal problem. Another nice property of linear programs is the simplicity of the construction of the dual problem from the primal. The construction of the dual can be explained by giving the following set of rules.

- The number of variables is equal to the number of constraints in the primal linear program.
- The coefficients of the objective function are the right-hand side coefficients of the constraints in the primal.
- The objective function is to be minimized.
- The number of constraints is equal to the number of variables in the primal linear program.
- The coefficients of the constraint corresponding to the variable x_i are the coefficients of this variable in each constraint of the primal.
- All the constraints are of the form $\geq$.

Example 3. As an example of duality, let us continue from Example 1. With the above transformations, we obtain the following dual linear program.

$$
\begin{array}{lrcrcrcrcl}
\text{Min} & & 15y_2 & + & 2y_3 & + & 8y_4 & & \\
\text{s.t.} & y_1 & - & y_2 & + & y_3 & + & y_4 & \geq & 1 \\
& -5y_1 & + & 5y_2 & - & y_3 & + & y_4 & \geq & 2 \\
& & & & & & & y_{1,2,3,4} & \geq & 0.
\end{array}
$$

The importance of the duality in the context of this chapter is the algorithmic implications that follow from this concept. Namely, because of the duality theorem, to reduce the run times the solver can decide whether to solve the primal or the dual problem. The reason for the speedup is the existence of the revised simplex algorithm, which utilizes only a matrix of the size proportional to the square of the number of constraints. Having this option, we can always choose the problem (primal or dual) with the smallest number of constraints. In Section 5, we will be dealing only with problems that have the same number of variables as there are constraints, which makes also the dual problem with the same properties. This further reduces the advantages of the revised simplex method and is a good motivation for the use of the basic simplex method which will be described next.

3.4 The Simplex Algorithm

Besides being a simple modeling tool in so many disciplines, the main reason for the success of LP is its elegance and efficiency in finding optimal solutions for highly complex problem instances.

There are many different variations of this fundamental algorithm. In what follows we will describe one variation, omitting all the proofs and details which are not relevant for the remainder of this chapter, but giving a detailed enough overview of the algorithm. More exhaustive descriptions of the algorithm can be found in many basic algorithm textbooks, e.g., Ref. [19].

We have already given a short description of the standard formulation of a linear program. Usually the problem is given in a matrix form in the following way. The n variables are described as a vector

$$x = [x_1, x_2, \ldots, x_n]^T.$$

The m constraints are defined by the coefficients of the linear combinations and given in the matrix form as:

$$Ax \leq b.$$

The objective function which we wish to minimize is given as an n-dimensional vector c of coefficients. The objective function is thus in vector form

$$cx.$$

Linear programs belong to a wider class of optimization problems, known as convex optimization problems [20]. On convex optimization problems local optimization algorithms always converge to the optimal solution. And the simplex algorithm is such a local optimization method that moves between different feasible solutions, until it reaches the local optimum and thus the optimal solution.

First, let us discuss the data organization during the algorithm. The simplex algorithm is usually described in terms of the simplex tableau, i.e., a tabular organization of the input data to the problem. This tableau contains the constraint matrix A, the first row containing the coefficients of the objective functions, the first column of the right-hand side coefficients of the constraints, and the first cell containing the current value of the objective function. Table 1 shows the layout of the data in a more graphical manner. This tableau will be referred as T.

With the data organized in the tableau form, we can describe the algorithm as a series of transformations of T, each transformed tableau representing a new feasible solution, with a better goal value than the solution on the previous step. The move from one feasible solution to the next feasible solution is characterized by a choice of a column and a row in T. The element at the intersection of the chosen column and row is called a pivot, and the chosen row and column are called pivot row and pivot column, respectively.

The choice of the pivot column is the most important factor for the speed of convergence to the optimal solution, and many different criteria exist [21]. These rules are mainly heuristics without any guarantees for the

Table 1 The Layout of the Linear Program as a Simplex Tableau T

$-cx$	c
b	A

The matrix A representing the coefficients of the constraints, vector c the coefficients of the objective function, vector b the right-hand side of the constraints, and cx is the current value of the objective function.

performance on specific instances. The goal of this chapter is not to investigate the efficiency of these rules, but the speedup a single iteration of the simplex method, so we chose to implement one of the rules, i.e., the choice of the pivot column j_p which has the largest coefficient in the objective function $c[j_p]$. This rule is usually referred to as the Dantzig rule.

The choice of the pivot column represents a choice of a variable which is going to increase the value of the solution. The amount of the increase is determined by the most strict constraint in A, which is the row i_p that has the minimal positive ratio

$$\frac{b[i_p]}{T[i_p, j_p]}.$$

After the choice of the pivot column and row have been made, the entire tableau is recomputed. This recomputation of T moves the algorithm to a new feasible solution with a better goal value. This transformation changes every element of the tableau according to four cases:

The pivot element

$$T[i_p, j_p] = -\frac{1}{p}$$

The pivot row

$$T[i_p, j] = \frac{T[i_p, j]}{p} \quad \forall j$$

The pivot column

$$T[i, j_p] = \frac{T[i, j_p]}{p} \quad \forall i$$

Other elements

$$T[i, j] = T[i, j] - row[j] \times col[i] \times \frac{1}{p} \quad \forall i \neq i_p, j \neq j_p,$$

where *row* is the old pivot row and *col* is the old pivot column.

The complete algorithm is given in Fig. 3.

$$
\begin{array}{ll}
\textbf{Data}: \text{Simplex tableau } T \\
1 \quad \textbf{while } \exists j : c[j] > 0 \textbf{ do} \\
2 \quad\quad j_p = \arg\max_j c[j]; \\
3 \quad\quad i_p = \arg\min_{i, T[i,j_p] > 0} \frac{b[i]}{T[i,j_p]}; \\
4 \quad\quad p = T[i_p, j_p]; \\
5 \quad\quad row = T[i_p, *]; \\
6 \quad\quad col = T[*, j_p]; \\
7 \quad\quad T[i,j] = \begin{cases} -\frac{1}{T[i,j]} & i = i_p \wedge j = j_p \\ \frac{p}{T[i,j]} & i = i_p \\ -\frac{p}{T[i,j]} & j = j_p \\ T[i,j] - row[j] \times col[i] \times \frac{1}{p} & \text{otherwise} \end{cases} \\
8 \quad \textbf{end}
\end{array}
$$

Fig. 3 The basic structure of the simplex algorithm.

4. ACCELERATED SIMPLEX ALGORITHM

In this section we describe various approaches to porting the simplex algorithm to the data-flow computer architecture. Namely, we present two approaches: (a) the data representing the simplex tableau are stored in the main memory and streamed to the data-flow unit for each pivoting operation, and (b) the data are stored in the special on-chip large memory and directly used by the data-flow unit. We also discuss a way to additionally parallelize these algorithms.

4.1 Algorithm Engineering

To adapt the simplex algorithm for the data-flow architecture we largely follow the algorithm engineering process, which, in general, refers to a process required to transform a pencil-and-paper algorithm into a robust, efficient, well-tested, and easily usable implementation [22]. An important part of the process is also the experimental evaluation through which practical efficiency and usability of the algorithm is established. For more details on the process see Refs. [23,24].

Here we briefly describe the process and its main phases, which are algorithm design, algorithm analysis, implementation, and experimental evaluation. See also Fig. 4 for a diagram representing the process. The phases generally follow each other, however, skipping, revising, and repeating a phase is a possibility, especially when multiple iterations are performed.

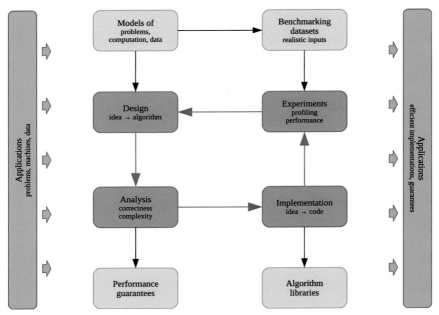

Fig. 4 Algorithm engineering process.

The initial phase is the algorithm design, where a designer, based on a model of computation (in our case the data-flow architecture and stream processing) constructs a pencil-and-paper algorithm for the corresponding problem (e.g., LP). She may also specify a particular model of data (e.g., dense simplex tableau). Our work is mainly focused on an efficient data choreography and interconnection of data-flow kernels.

The outcome of the algorithm analysis phase is the correctness proofs and computational complexity estimations as well as performance guarantees. Since the simplex algorithm is a well-known algorithm these results are readily available [13]. Hence, we mostly skipped this phase.

Our main effort was put in the implementation phase, where we implemented several variants of the simplex algorithm employing various data-flow programming techniques. Most of the following sections deal with this phase.

The final phase of the algorithm engineering process is the phase of experimental evaluation, whose goals is to establish practical efficiency of the proposed algorithms. We describe this phase in details in Section 5.

4.2 Streaming From the Main Memory

In this section we explore the acceleration approach, where the simplex tableau data are kept in the main memory and for each pivoting operation the whole tableau is streamed to the data-flow processor. All other operations, such as the choice of column and the selection of row are performed by the CPU as they are significantly less computationally intensive (by an order of magnitude). Since the tableau is present in the main memory, the CPU has all the required data and, consequently, all the control over how to proceed or when to stop the algorithm.

The pivoting operation requires the simplex tableau and two vectors, i.e., the pivot column and pivot row. These three objects are given to the DFE as input streams, which are described in the following list.

- The simplex tableau is streamed sequentially row by row.
- The streaming of pivot column is performed steadily by reading in only one value from stream per each row of tableau.
- The pivot row is streamed in while reading in the first row of tableau and stored to the on-chip FMem, whereas for subsequent rows of tableau reading from the FMem is rather performed instead.

The result of pivoting operation, i.e., the entire recomputed simplex tableau, is then streamed back to the main memory as the kernel output. See also Fig. 5 for schematic overview of data movements.

The data-flow kernel of the pivoting operation is shown in Fig. 6. The corresponding MaxJ class is named **PivotingKernel** and its constructor contains the kernel implementation. The argument **fmemDepth** of the constructor specifies the maximum length of one row of tableau. In lines 10–14, the scalar inputs are first read. In lines 16–19, counters and control variables are defined. The on-chip FMem is allocated and initialized in lines 21–23. The

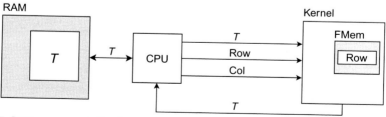

Fig. 5 Data movements for the approach streaming from the host. The simplex tableau T is stored in the main memory. The tableau and the selected pivot row and column are streamed to the kernel.

```
 1   class PivotingKernel extends Kernel {
 2       public static final String NAME = PivotingKernel.class.getSimpleName();
 3       static final DFEType dataType = dfeFloat(8, 24);
 4
 5       PivotingKernel(KernelParameters parameters, int fmemDepth) {
 6           super(parameters);
 7           final int bitCnt = MathUtils.bitsToAddress(fmemDepth);
 8           final DFEType addrType = dfeUInt(bitCnt);
 9
10           DFEVar m = io.scalarInput("m", addrType);
11           DFEVar n = io.scalarInput("n", addrType);
12           DFEVar row = io.scalarInput("row", addrType);
13           DFEVar col = io.scalarInput("col", addrType);
14           DFEVar pivot = io.scalarInput("pivot", dataType);
15
16           CounterChain chain = control.count.makeCounterChain();
17           DFEVar i = chain.addCounter(m, 1);
18           DFEVar j = chain.addCounter(n, 1);
19           DFEVar firstrow = i === 0;
20
21           Memory<DFEVar> pivrow = mem.alloc(dataType, fmemDepth);
22           DFEVar pivrowstream = io.input("pivrow", dataType, firstrow);
23           pivrow.write(j, pivrowstream, firstrow);
24
25           DFEVar pivcol_i = io.input("pivcol", dataType, j === 0);
26           DFEVar pivrow_j = firstrow ? pivrowstream : pivrow.read(j);
27
28           DFEVar x = io.input("x", dataType);
29           DFEVar t1 = x * pivot;
30           DFEVar t2 = x - pivot * pivcol_i * pivrow_j;
31           DFEVar isrow = i === row;
32           DFEVar iscol = j === col;
33           DFEVar select = iscol.cat(isrow);
34           DFEVar y = control.mux(select, t2, t1, -t1, pivot);
35           io.output("y", y, dataType);
36       }
37   }
```

Fig. 6 Class PivotingKernel in the MaxJ programming language implementing the pivoting operation of the host-streamed approach.

variable **pivcol_i** contains the i-th element of pivot column, which is read in when processing first column. Similarly, **pivrow_j** contains the j-th element of pivot row (lines 25 and 26). The actual pivoting operation of one tableau element is implemented in lines 28–35. Notice also that the pivot scalar input is actually its inverse (calculated by control-flow part).

4.3 Concurrent Streaming From the Main Memory

The main drawback of this approach is that the number of operations per one tableau element is too low, and thus not exploiting the full potential of the possible DFE parallelism. Additionally, due to technological constraints, i.e., low clock frequency, the data-flow implementation of an algorithm can be significantly faster than the CPU implementation only if massively parallelized.

One approach to parallelization is to increase the pipeline depth. Unfortunately, this approach may not be fruitful for the case of the simplex algorithm for the above described reason. Another approach is the introduction of multiple *pipes*, i.e., many parallel streams concurrently fed into the DFE. In what follows we describe how to introduce multiple pipes into the basic data-flow algorithm.

To introduce pipe-based concurrency, we first replace the constructor declaration of **PivotingKernel** class with

```
PivotingKernel(KernelParameters parameters, int fmemDepth,
    int vectorDepth),
```

where the parameter **vectorDepth** specifies the number of pipes. Then, inside the constructor, we define new vector data type

```
final DFEVectorType<DFEVar> vecType =
    new DFEVectorType<DFEVar>(dataType, vectorDepth);
```

which represents vectors of length **vectorDepth** containing elements of type **dataType**.

Fortunately, the pivoting operation recomputes individual elements of the tableau completely independently and is, therefore, very amenable to parallelization, even to the point where each element of the tableau is processed in its own pipe. In theory the problem is embarrassingly parallel, however, in practice it early encounters a bottleneck of insufficient data transfer rates. The main drawback turns out to be low throughput of the bus between the CPU and the DFE. The exact implications of this are described in Section 5, where the empirical tests will clearly show the limits of this approach.

Now to finish the parallelization we employ the MaxJ's built-in **DFEVector** construct which greatly simplifies the transformation. We introduce vector counter by replacing line 18 with

```
DFEVector<DFEVar> j = chain.addCounterVectNoWarn
    (vectorDepth, n, 1);
```

Additionally, we also replace lines 21–35 with

```
Memory<DFEVector<DFEVar>> pivrow = mem.
    alloc(vecType, fmemDepth);
DFEVector<DFEVar> pivrowstream = io.input
    ("pivrow", vecType, firstrow);
pivrow.write(j[0], pivrowstream, firstrow);

DFEVar pivcol_i = io.input("pivcol", dataType, j[0] === 0);
DFEVector<DFEVar> pivrow_j = firstrow ? pivrowstream :
    pivrow.read(j[0]);

DFEVector<DFEVar> x = io.input("x", vecType);
DFEVector<DFEVar> t1 = x * pivot;
DFEVector<DFEVar> t2 = x - pivot * pivcol_i * pivrow_j;
DFEVar isrow = i === row;
DFEVector<DFEVar> iscol = j === col;
DFEVector<DFEVar> select = iscol.cat(isrow);
DFEVector<DFEVar> pivots = DFEVectorType.
    newInstance(vectorDepth, pivot);
DFEVector<DFEVar> y = control.mux(select, t2, t1, -t1, pivots);
io.output("y", y, vecType);
```

In the above lines, **DFEVector** type is utilized: classical arithmetic operations, e.g., multiplication, subtraction, are vectorized in an element-wise manner.

4.4 Streaming From the On-chip Large Memory

To mitigate the aforementioned bottleneck the data has to be moved to the on-chip large memory (LMem), which has a much higher bandwidth than the peripheral component interconnect express (PCIe) bus. Since the large memory differs from the main memory in the way an efficient access to the data can be made, other modifications, besides moving the data closer to the DFE, are necessary. Due to this, a significant change also has to be made in the implementation.

First let us have a look at the code changes for the control-flow part of the algorithm. During the execution it does not have all the data, but just enough to control the whole process. See Fig. 7 for the algorithm overview. The following list exposes the main changes to the basic algorithm.

• Before the start of the algorithm the entire simplex tableau must be transferred to the LMem.

Data: Simplex tableau T
1   $write\_tableau\_to\_lmem(T)$;
2   $col = select\_col(T)$;
3   $row = select\_row(T)$;
4   **while** $col[0] > 0$ **do**
5      $pivoting\_DFE(col, row, out\_b, out\_col)$;
6      $row\_idx = select\_row(out\_b, out\_col)$;
7      $row = read\_row\_from\_lmem(row\_idx)$;
8      $col = out\_col$;
9   **end**

Fig. 7 Source code of the control part for the LMem-streamed algorithm.

- Besides the pivoting operation, the DFE part now performs also the pivot column selection (except for the first iteration, when this is still done by the CPU).
- To perform the row selection the CPU needs to access the pivot column and the bounds column. Hence, these two are outputs of DFE.

Observe also, that the first element of the pivot column is actually the coefficient of the objective function; when it becomes negative, the optimal solution has been found and the algorithm stops. See also Fig. 8 for an overview of the data movements between CPU and DFE.

The driving force for these changes is the fact that the large memory provides a fast access only to a linear block of addresses. As the simplex tableau can be linearized either by rows or by columns, one must opt for fast access of either rows or columns. We chose the former and, consequently, any row can be extracted easily from the large memory but extracting the pivot column demands processing the whole tableau.

In order to be able to achieve this, we split the functionality of the DFE into three different kernels.

Pivoting kernel This kernel is mostly the same as the one in the host-streamed approach except that the tableau streaming is now done from the large memory instead of from the CPU. However, the pivot column and the pivot row are still streamed from the CPU. See also Fig. 6 for a reference: the only important addition here is that tableau is actually output twice. Additionally, the objective function is also a part of the output. The corresponding data-flow graph is depicted in Fig. 9.

Selection kernel The input to this kernel, i.e., the vector c, comes from the pivoting kernel. Then, the maximum positive value is found and its index is returned to the filtering kernel. The algorithm source code may

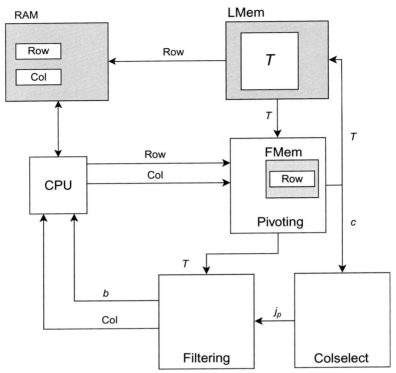

Fig. 8 Data movements for the approach streaming from the on-chip large memory. The simplex tableau *T* is stored in the large memory. Only pivot row and column are streamed to the kernel from the CPU.

be found in Fig. 10. In the implementation two counters are used, i.e., the first, *i* (line 15), is used as the index of the input, and the second, *l* (line 16), is used as a counter to accurately carry out the graph loop made by the variable **oldmaxval**. To reduce the generated loop length, an architecture-specific optimization is used (lines 20 and 25). Notice also, in line 26, how the special programmatic construct **streamHold** is used to implement the maximum. This construct can be replaced with a multiplexer if needed. See also the corresponding data-flow graph in Fig. 12.

Filtering kernel This kernel takes care of filtering the pivot column and bound ratios out of the simplex tableau, which is streamed in as the input. The index of the pivot column comes in from the selection kernel. The filtering source code is available in Fig. 11.

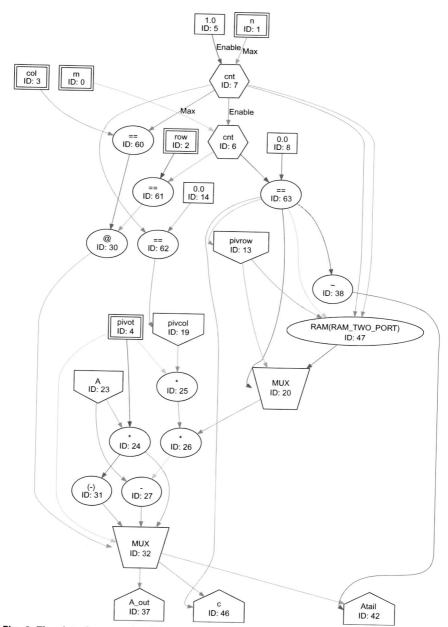

Fig. 9 The data-flow graph of the pivoting kernel for the LMem-streamed algorithm.

```
 1   public class SelectionKernel extends Kernel {
 2       public static final String NAME = SelectionKernel.class.getSimpleName();
 3           static final DFEType dataType = dfeFloat(8, 24);
 4           static final DFEType idxType = dfeUInt(32);
 5
 6       protected SelectionKernel(KernelParameters parameters) {
 7           super(parameters);
 8
 9           OffsetExpr loopLength = stream.makeOffsetAutoLoop("loopLength");
10           DFEVar loopLengthVal = loopLength.getDFEVar(this, dfeUInt(8));
11
12           DFEVar n = io.scalarInput("n", idxType);
13
14           CounterChain chain = control.count.makeCounterChain();
15           DFEVar i = chain.addCounter(n, 1);
16           DFEVar l = chain.addCounter(loopLengthVal, 1);
17
18           DFEVar c = io.input("c", dataType, l === 0);
19
20           optimization.pushPipeliningFactor(0);
21           DFEVar oldmaxval = dataType.newInstance(this);
22           DFEVar greater = c > oldmaxval;
23           DFEVar maxval = Reductions.streamHold(c, greater);
24           oldmaxval <== stream.offset(maxval, -loopLength);
25           optimization.popPipeliningFactor();
26
27           DFEVar maxidx = Reductions.streamHold(i, greater);
28
29           io.output("c_out", c, dataType, l === 0);
30           io.scalarOutput("maxcol_out", maxidx, idxType);
31           io.output("maxcol", maxidx, idxType, i === n - 1);
32       }
33   }
```

Fig. 10 Class SelectionKernel implementing the operation of the pivot column selection.

The output of the filtering kernel is then used by the CPU to select the pivot column only to be used in the next iteration. The CPU also extracts the corresponding row from the large memory and feeds it to the pivoting kernel. This is the main reason for the column select kernel which reads as input the goal function coefficients c and selects (e.g., maximal positive) coefficient, i.e., the column. This column as well as the bounds column are then filtered from the tableau only to be streamed to the CPU. The CPU then calculates the index of the pivot row and extracts it from the large memory only to be used in the next iteration.

4.5 Concurrent Streaming From the On-chip Large Memory

Introducing the pipe-based concurrency to LMem-streamed algorithms is similar as for the host-streamed algorithms. Thus, we keep this section brief

```
 1   public class FilteringKernel extends Kernel {
 2       public static final String NAME = FilteringKernel.class.getSimpleName();
 3       static final DFEType dataType = dfeFloat(8, 24);
 4       static final DFEType idxType = dfeUInt(32);
 5
 6       protected FilteringKernel(KernelParameters parameters, int vectorDepth) {
 7           super(parameters);
 8
 9           final DFEVectorType<DFEVar> vecType = new
                 DFEVectorType<DFEVar>(dataType, vectorDepth);
10
11           DFEVar n = io.scalarInput("n", idxType);
12
13           DFEVar tick = control.count.simpleCounter(32);
14           DFEVar first = control.count.pulse(1); // tick === 0;
15           DFEVar firstrow = tick < n / vectorDepth;
16           DFEVector<DFEVar> j = control.count.simpleCounterVect(vectorDepth, 32,
                 n);
17
18           DFEVar col = io.input("maxcol", idxType, first);
19           DFEVector<DFEVar> A = io.input("Atail", vecType, ~firstrow);
20           DFEVar b = Reductions.streamHold(A[0], j[0] === 0);
21
22           DFEVector<DFEVar> iscol = j === col;
23           DFEVar select = iscol.pack().cast(dfeUInt(vectorDepth));
24           DFEVar AA = control.oneHotMux(select, A.packToList());
25
26           DFEVar ratio = firstrow ? 0 : b / AA;
27           DFEVar pivcol = firstrow ? 0 : AA;
28           DFEVar fire = select > 0;
29           io.output("ratio_out", ratio, dataType, fire);
30           io.output("pivcol_out", pivcol, dataType, fire);
31       }
32   }
```

Fig. 11 Class FilteringKernel implementing the operation of filtering out the pivot column and bounds.

and state only the differences. In general, three kernels may be parallelized here, but we choose only to parallelize the two: the pivoting kernel and the filtering kernel.

The reason behind not parallelizing the selection kernel is that it runs while the pivoting kernel also runs. Additionally, the latter runs much longer than the former for it processes the entire tableau as opposed to the selection which scans only one row. Thus, the selection is finished long before the pivoting (despite the loop in the data-flow graph, see Fig. 12).

As the simplex tableau is now placed in the large memory, we managed to obtain a much larger parallelization factor (and consequently acceleration) than with the data streamed from the main memory. The exact speedup factors and a careful empirical evaluation are described in the next section.

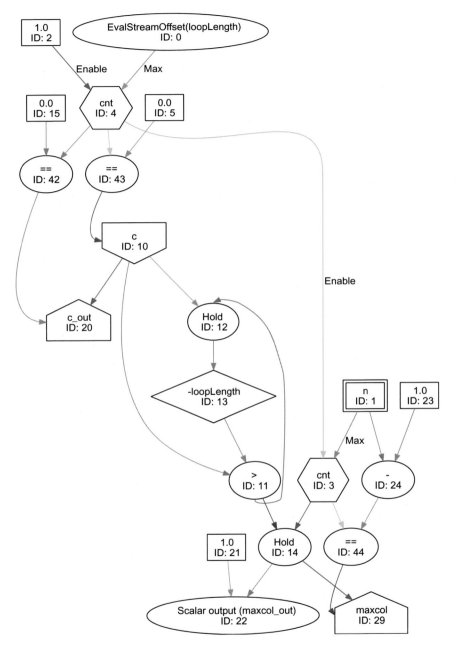

Fig. 12 The data-flow graph of the selection kernel (*left-hand side*) and filtering kernel (*right-hand side*) for the simplex algorithm streaming from the large memory.

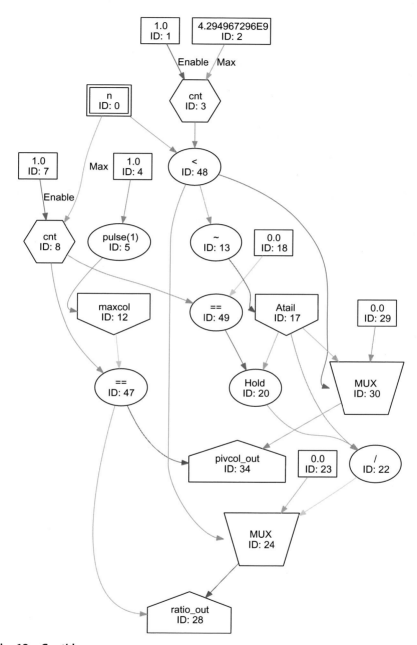

Fig. 12—Cont'd

5. EXPERIMENTAL EVALUATION

In this section we present our results of empirical comparison of the above described algorithms. Each comparison includes the CPU-only implementation for a reference as well as a brief discussion. We split the experimental evaluation into several categories. First, we compare different versions of host-streamed approach varying in the number of pipes. Second, we make similar comparison of LMem-streamed approach variations. Third, we compare the algorithm performance on different densities of simplex tableau. Fourth, we present a summary of evaluation, and, finally, we discuss FPGA resource usage. But before we begin, let us present the benchmark configuration used in the experiments.

5.1 Experimental Goals

One of the most important planning steps, which are necessary for good experimental research, is to define experimental goals [23,24]. The experimenter needs to think about and define the motivation for the experiments, questions needing answers, statements needed to be verified, and newsworthiness of the experiment.

The main goal of our work is, besides showing that it is possible to transfer the classical simplex algorithm from the control-flow architecture into the data-flow environment, to demonstrate that the data-flow variant is also competitive in a small to moderate size data-flow configuration. As shown in the following sections this goal was completely achieved.

Another important goal of our experimentation is to rank different variants of data-flow algorithms according to various engineering techniques used. In particular, we are mainly interested in the speedups which are attainable via usage of the data-flow engine large memory and via usage of different number of pipes.

Furthermore, different variants of the algorithm consume different amounts of FPGA resources. Hence, our goal is also to study how the resource consumption depend on the increase of data-flow engine-specific parameters, i.e., number of pipes, and what is the maximum value of the parameters that still allows for a successful compilation of the data-flow graph.

Certainly, our goal is to determine the best algorithm out of all the proposed ones and to provide an experimental basis for ranking the algorithms. We are interested in running time of the algorithms as well as the maximum possible acceleration (in comparison to the plain control-flow version) that could be achieved using the data-flow platform.

Importantly, with our last experimental study, we explored the applicability of the approach in practice. Since the classical simplex algorithm is more suitable for dense simplex tableaux we experimentally determined the tableau density level where it is better to use data-flow version than the out-of-the-box LP solver.

Finally, the importance of the above listed goals forms a firm basis for the newsworthiness of our experiments.

5.2 Benchmark Configuration

Machine Configuration: To perform the empirical evaluation of the above described algorithms we utilized a high-performance data-flow computing platform provided by Maxeler Technologies, Ltd. The platform contains control-flow as well as data-flow computing units. The machine configuration is the following.

- The control-flow part consists of the Intel Xeon X5650 multicore-architecture running at 2.67 GHz with three levels of cache (L1: 32 kB + 32 kB, L2: 256 kB, and L3: 12 kB) and 48 GB of memory.
- The platform is supplemented with two data-flow coprocessors situated on MAX3424A (revision 07) cards where reconfigurable part consists of Xilinx Virtex-6 SX475T (XC6VSX475T) FPGA chip with 297,600 lookup tables (LUTs), 595,200 flip flops (FFs), 1064 dual port 36 kbit block random access memories (BRAMs), and 2016 DSPs. The card communicates with the host via PCIe bus with a bandwidth of 2 GB/s.

Test Set: Benchmark test set consists of 10 test cases each of different size $n \times n$ of the simplex tableau where n varies from 1000 to 10,000 in steps of 1000. The density of generated tableaux is 100% unless specified otherwise. Henceforth, the tableaux generation algorithm is straightforward. To generate a tableau with a given density ρ we fill the column b and the row c in with all positive values as well as the A matrix until the total number of non-zero elements is at most ρ percent of n^2.

All versions of the algorithms were run on the above test set. When an algorithm finishes it reports the objective value achieved, the number of pivoting operations, and the elapsed wall-clock time. Notice that, the density as well as the structure of linear program has no effect on our benchmarks, since

- the running time of the pivoting operation does not depend on the tableau density (we also demonstrate density independence by an experiment), and
- each algorithm was programmed to stop after at most 10,000 iterations.

Such a benchmark basically provides comparison based on an average time per an iteration of simplex algorithm (i.e., if the total time is divided by the number of iterations). For comparison we also include CPU-only implementation in order to explore the potential of various acceleration techniques.

Using the obtained results on running times of particular implementations we calculate the speedup, respectively, acceleration factor as

$$accel = \frac{t_{CPU}}{t_{DFE}},$$

where t_{CPU} is the running time of the CPU-only algorithm, and t_{DFE} is the running time of a particular DFE-based algorithm. These factors are also shown in the following sections as graphical plots depending on the problem size. In addition, we also calculate an average acceleration factor for each algorithm.

5.3 Host-Streamed Algorithms

First, let us examine the host-streamed data-flow versions of the classical simplex algorithm (described in Section 4.2). We implemented and evaluated several parallel versions varying in the number of concurrently processed elements, i.e., pipes. Namely, we tested the implementations with no parallelism (host-none), two pipes (host-pipe2), four pipes (host-pipe4), and eight pipes (host-pipe8). Since the input stream length in bytes must be divisible by 16 which corresponds to 4 floating-point numbers, the concurrency level must also be a factor or multiplier of 4. See Table 2 for the

Table 2 Summary of Host-Streamed Implementations

Algorithm	Freq	Cols	Pipes
host-none	200	12,288	1
host-pipe2	200	12,288	2
host-pipe4	200	12,288	4
host-pipe8	200	12,288	8

The column *freq* gives the stream clock frequency in MHz, the *cols* gives the maximum number of columns of simplex tableau, and *pipes* gives the concurrency level in terms of number of pipes.

summary of implemented host–streamed algorithms. Notice that in all cases the compiler was able to synthesize the FPGA circuit with the best possible frequency setting while not producing any timing problems.

In all the implementations the stream clock frequency was set to 200 MHz. Using also the bandwidth of PCIe, i.e., 2 GB/s, we estimate the upper bound on bytes processed per clock tick to 10 bytes/tick. Henceforth, since single–precision floating-point number representation requires 4 bytes, the speedup of concurrent host–streamed processing may be at most 2.5 over **host-none** (as opposed to the naïvely expected 4 for **host-pipe4**) as is also confirmed by our experiments. Consequently, the PCIe bandwidth represents a major bottleneck and its improvement would significantly boost the algorithm performance.

The results of the experiments, i.e., the running times and the achieved acceleration factors (compared to the CPU-only implementation) are plotted in Figs. 13 and 14, respectively. Observe that acceleration factors stabilize when the input size grows. Hence, we calculated the average acceleration factors only for inputs of size $n \times n$, where $n \geq 5000$. See Table 3 for the comparison. See also Table 4 for an overview of the FPGA resource utilization as well as Fig. 19 for its graphical representation. Here we list some of our observations:

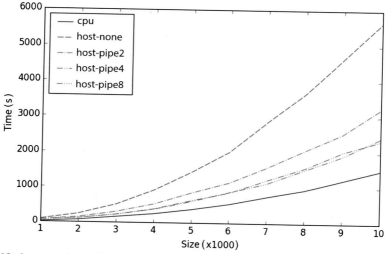

Fig. 13 A comparison of running times for the host-streamed implementations. Times are expressed in terms of simplex tableau size 1000n × 1000n.

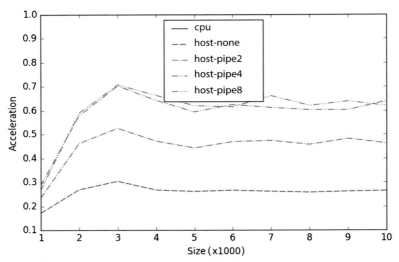

Fig. 14 Acceleration factors of host-streamed implementations. Factors are expressed in terms of simplex tableau size 1000n × 1000n.

Table 3 Average Acceleration Factors for Host-Streamed Implementations

Algorithm	**vs** cpu	**vs** Host-None
cpu	1.0	3.81
host-none	0.26	1.0
host-pipe2	0.47	1.79
host-pipe4	0.62	2.35
host-pipe8	0.63	2.41

The factors are calculated only in the respect of test sets where $n > 5000$. The second and third columns give acceleration factors over cpu and **Host-None**, respectively.

Table 4 Resource Utilization [%] for Host-Streamed Implementations

Algorithm	LUTs	FFs	BRAMs	DSPs
host-none	2.32	1.42	2.63	0.30
host-pipe2	2.61	1.60	3.76	0.50
host-pipe4	3.11	1.87	5.45	0.89
host-pipe8	4.31	2.74	10.62	1.69

Notice that, **Host-Pipe8** is not significantly faster than **Host-Pipe4**. Thus, the extra spent resources are not worth the effort.

- None of the host-streamed versions is faster than the CPU-only implementation. The reason behind this is twofold: (1) low PCIe throughput and (2) excessive transferring of the simplex tableau between the main memory and the data-flow unit.
- The nonconcurrent version (**host-none**) obtains an average acceleration factor of 0.26 (calculated for $n \geq 5000$), i.e., it is more than times slower than the CPU-only version.
- The concurrent version using two pipes, i.e., **host-pipe2**, obtains an average acceleration factor of 0.47 over **cpu** and 1.79 over **host-none**, where the upper bound for the latter is 2.
- Better performance, but still unsatisfiable, is obtained with either **host-pipe4** or **host-pipe8** where the latter is not significantly faster than the former. The average measured acceleration for both is approximately 0.62. For the reasons already explained above the average acceleration over **host-none** cannot be better than 2.5; indeed, it is 2.35 and 2.41, respectively. Consequently, the estimated upper bound over **cpu** is about $2.5 \times 0.26 = 0.65$, which strongly supports the measured one 0.62.
- We consider **host-pipe4** as the most efficient version. Using more pipes currently does not seem to significantly pay off.
- Concerning the FPGA resources used for each implementation their utilization is very low. In particular, the most expended resource is BRAM used for the synthesis of on-chip FMem; its consumption is less than 5% of the total available for **host-pipe4**.
- The most economical algorithm in terms of speed up vs resources spent is **host-pipe4**.

In our implementations the depth of FMem is set to 12 K of floating-point numbers. Considering **host-pipe4** such depth uses about 5% of BRAMs. Hence, in principle, an order of magnitude of increase of FMem is possible; we confirmed this also by an experiment: we were able to easily compile a program with 100 K of numbers (about 25% of BRAMs used). Since FMem is used to store one row of simplex tableau, linear programs with 100 K variables could be solved. Notice that storing programs of size $(100 \text{ K})^2$ would already use up to about 40 GB of main memory.

Even though, none of the host-streamed data-flow implementations turned out to be better than CPU-only approach we still see their potential for the future. Currently, the data-flow-based acceleration card supports only 2 GB/s data throughput via PCIe bus. However, there are already much better throughputs available in upgrades of PCIe, i.e., the announced PCIe version 4.0 supports up to 31.508 GB/s [25].

5.4 LMem-Streamed Algorithms

Now, let us focus on LMem–streamed data–flow versions of the simplex algorithm (described in Section 4.4). We evaluated implementations differing in the number of pipes ranging from no concurrency (lmem-none) to 48-fold concurrency (namely, lmem-pipe2, lmem-pipe4 lmem-pipe8, lmem-pipe12, lmem-pipe16, lmem-pipe24, and lmem-pipe48).

Similarly to the host-streamed approach the same restrictions apply for the concurrency level. Additionally, the concurrency level must also be a factor or multiple of 48 which is the burst size of the bus between the large memory and the DFE, (i.e., 192 bytes or 48 single-precision floating-point numbers).

In all the implementations the stream clock frequency was set to 150 MHz while the frequency of large memory was set to 333 MHz. The compiler failed to place and route the design on the chip with higher frequencies, in order to satisfy the timing constraints. See Table 5 for the overview of the implemented host-streamed algorithms.

To estimate the upper bound on the acceleration factor we proceed by dividing the bandwidth (32 GB/s) of large memory bus with the stream frequency (150 MHz) and the size 4 bytes for representation of floating-point numbers as well as 2 (bidirectional memory access). Doing this we get about 26 as an upper bound for the speedup over nonconcurrent implementation. Notice that the actual acceleration of lmem-pipe24 over lmem-none is around 19.6.

Table 5 Summary of LMem-Streamed Implementations

Algorithm	Freq	LMem	Cols	Pipes
lmem-none	150	333	12,288	1
lmem-pipe2	150	333	12,288	2
lmem-pipe4	150	333	12,288	4
lmem-pipe8	150	333	12,288	8
lmem-pipe12	150	333	12,288	12
lmem-pipe16	150	333	12,288	16
lmem-pipe24	150	333	12,288	24
lmem-pipe48	150	333	12,288	48

The column *freq* gives the stream clock frequency in MHz, the *LMem* gives the large memory frequency in MHz, the *cols* gives the maximum number of columns of simplex tableau, and *pipes* gives the concurrency level in terms of number of pipes.

The results of experimental evaluation of LMem–streamed algorithms are shown in Fig. 15 (running times) and Fig. 16 (accelerations). The average acceleration factors are given in Table 7. See also Table 6 for an overview of the FPGA resource utilization as well as Fig. 19 for its graphical representation. The following list summarizes our observations.

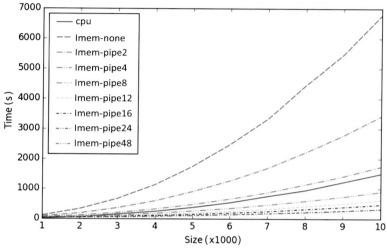

Fig. 15 Comparison of running times of LMem-streamed implementations. Times are expressed in terms of simplex tableau size $1000n \times 1000n$.

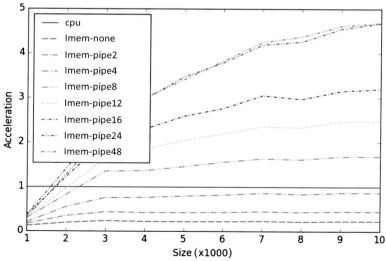

Fig. 16 Acceleration factors of LMem-streamed implementations. Factors are expressed in terms of simplex tableau size $1000n \times 1000n$.

Table 6 Resource Utilization [%] for LMem-Streamed Implementations

Algorithm	LUTs	FFs	BRAMs	DSPs
lmem	12.68	9.74	14.19	0.30
lmem-pipe2	13.14	9.97	15.51	0.50
lmem-pipe4	13.50	10.31	17.76	0.89
lmem-pipe8	14.68	11.14	22.93	1.69
lmem-pipe12	16.27	12.00	27.63	2.48
lmem-pipe16	16.88	12.66	32.61	3.27
lmem-pipe24	20.27	14.40	42.29	4.86
lmem-pipe48	28.32	18.98	59.21	9.62

Table 7 Average Acceleration Factors for Host-Streamed Implementations

Algorithm	vs cpu	vs LMem-none
cpu	1.0	4.75
lmem-none	0.22	1.0
lmem-pipe2	0.43	1.98
lmem-pipe4	0.85	3.86
lmem-pipe8	1.63	7.44
lmem-pipe12	2.36	10.76
lmem-pipe16	3.03	13.82
lmem-pipe24	4.29	19.59
lmem-pipe48	4.34	19.84

The factors are calculated only in the respect of test sets where $n > 5000$. The second and third columns give acceleration factors over cpu and Host-None, respectively.

- Nonconcurrent version lmem-none and up to four pipe concurrency, i.e., lmem-pipe2 and lmem-pipe4, were not able to outperform the CPU–only approach.
- Implementations using eight or more pipes provide better speedups over the cpu. In particular, average accelerations (where $n > 5000$) achieved by lmem-pipe8, lmem-pipe12, lmem-pipe16, lmem-pipe24, and lmem-pipe48 are 1.63, 2.36, 3.03, 4.29, and 4.34, respectively.
- In terms of speed up over lmem-none the average accelerations (where $n \geq 5000$) are up to almost 20. See Table 7 for details.

- Notice that, the acceleration factor is increasing proportionally to n. From the plots in Fig. 16 it seems that lmem-pipe24 and lmem-pipe48 acceleration factors could be even better with bigger instances. In particular, these two versions achieve acceleration factor (over cpu) 4.67 for $n = 10,000$, and 4.75 for $n = 12,000$ (not shown in the plot).
- Implementation lmem-pipe48 is not significantly better than lmem-pipe24. Hence, we consider the latter as the most efficient version.
- The most consumed FPGA resource is, similarly to host-stream approach, BRAM cells. Its consumption for lmem-pipe24 is about 42%.

Similarly as with host-streamed algorithm also here the depth of FMem is set to 12 K of floating-point numbers. Notice that, lmem-pipe24 uses about 43% of BRAMs. Hence, in principle, at least a double increase of FMem is possible, which would allow for linear programs with at least 20,000 variables to be solved with lmem-pipe24.

5.5 Summary

To summarize the results we compare the selected algorithms from each of the groups. In particular, we compare the CPU-only (i.e., cpu) implementation with the basic host- and LMem-streamed versions, i.e., host-none and lmem-none, as well as the most economic version (in terms of the speed up achieved and resources used), i.e., host-pipe4 and host-pipe24. The graphical comparison of running times is depicted in Fig. 17 and acceleration factors are shown in Fig. 18.

- Clearly, the fastest version is lmem-pipe24 which provides nearly fivefold speedup over cpu for big enough input tableaux.
- Straightforward host-streamed host-none and LMem-streamed lmem-none approaches are unable to outperform the CPU-only approach. Notice that lmem-none is slower than host-none because its stream clock frequency is lower.
- Version streamed from large memory lmem-pipe24 is better than host-pipe4 mostly because of the much better throughput of large memory.
- There are some initialization costs in the data-flow approaches, but acceleration factor stabilizes when input instances are getting bigger.
- Even though host-stream approaches are inferior in terms of speed we see their potential in low resource usage and the speedup when using wider PCIe bus.

Finally, let us present graphically (see Fig. 19 present resource utilization of all described algorithms). Observe that for the same number of pipes host-streamed approach utilizes less resources than a corresponding LMem-streamed one.

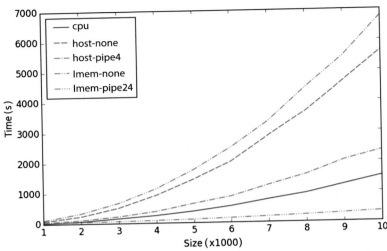

Fig. 17 Comparison of running times of the selected implementations. Times are expressed in terms of simplex tableau size $1000n \times 1000n$.

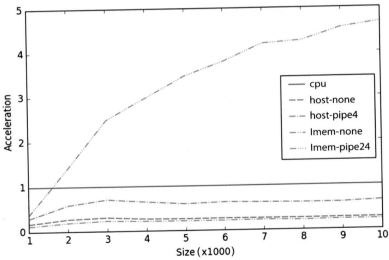

Fig. 18 Acceleration factors of the selected implementations. Factors are expressed in terms of simplex tableau size $1000n \times 1000n$.

5.6 Simplex Tableau Density

The pivoting operation in the classical simplex algorithm alters all elements (zero and nonzero ones) of the simplex tableau. Hence, its running time is independent from the tableau density, i.e., its asymptotic complexity is $O(n^2)$. On the other hand, publicly available LP solvers usually

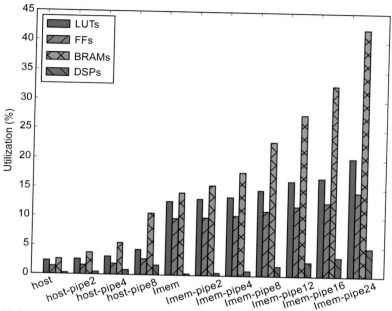

Fig. 19 Resource utilization per each algorithm.

exploit the sparseness of the tableau by using the revised simplex algorithm or some other even more suitable method. However, as it is shown by our experiments, even for moderately dense tableaux the classical simplex algorithm may be more efficient. In this section our goal is to explore at which densities it may be suitable to switch to the accelerated version.

For tableaux of size about 1000×1000 and less, the CPU-only algorithm is faster than any accelerated approach, but for sizes of 2000×2000 and above the presented accelerated versions begin to show their potential. This we also used in our experiments.

For the density benchmark we generated 10 test cases having simplex tableaux of varying density from 1% to 9%. We run and compare three algorithms:

- classical simplex CPU-only implementation denoted with **cpu**,
- LMem-streamed accelerated version with 24 pipes denoted with **lmem-pipe24**, and
- public available solver lp_solve [26] denoted with **lp_solve**.

Running times depending on the tableau density are plotted in Fig. 20. As in the previous sections, our goal was comparison based on the running time

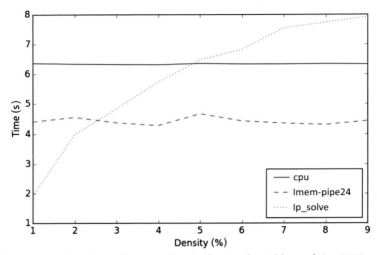

Fig. 20 Running time dependence on density for simplex tableau of size 2000 × 2000.

per pivoting operation. Thus, we stopped all algorithms after 10,000 iterations. However, **lp_solve** always finished its execution but its running time was normalized to 10,000 iterations.

There are two important observations.

- Running times of both (accelerated and nonaccelerated) classical simplex implementations are indeed independent from the tableau density. This is not the case for **lp_solve** algorithm which shows strong dependence on the tableau density.
- The density threshold, where the classical algorithm outperforms **lp_solve** is approximately 3% for the accelerated version and 5% for the nonaccelerated one.

Our experiment clearly shows that the classical simplex method is very competitive when the LP programs are not very sparse. Additionally, the density level where the acceleration becomes superior is relatively low, i.e., only a few percent.

Finally, it is informative to see the same experiment on bigger simplex tableaux of size 4000 × 4000. Here **lmem-pipe24** even outperforms **lp_solve** on tableaux of 1% density. See Fig. 21 for the details.

5.7 Discussion and Experimental Principles

According to the algorithm engineering discipline [23,24,27] there are several principles that should be followed in an experimental evaluation of

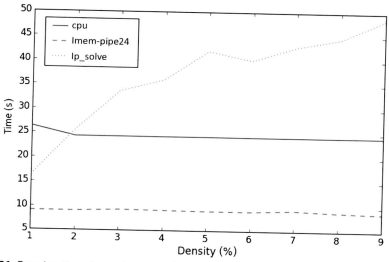

Fig. 21 Running time dependence on density for simplex tableau of size 4000 × 4000.

algorithms. In what follows we enlist these principles and discuss their particularities regarding our experimental considerations.

Reproducibility: Retaking the same experiment should produce similar results. The performance of data–flow algorithms is very predictable, since the data–flow unit is less susceptible to environmental or noise parameters (such as the load of computer system or the behavior of operating system), which cannot be explicitly manipulated. Additionally, running time of a data–flow algorithm depends solely on the input size, but not on the actual data in the input. The reason for this is the way the data–flow algorithms are written (i.e., no branch instructions).

We run all the experiments on the same computational platform and machine configuration specified in Section 5.2. Additionally, the source code of all the algorithms is available for download via the GitHub portal [28] as well as Maxeler's Application Gallery [29]. Furthermore, the generator program for the input test instances is included within the source-code packages.

Correctness: Indicators obtained from the experiment must accurately reflect the properties being studied. Performance indicators that are less influenced by external or noisy factors should be used.

In the experiments we measured wall-clock time, which is usually used in performance evaluation of parallel algorithms. During the experimentation the computer was used solely for this purpose. Notice also, that our research focus was on the simplex pivoting operation (and not on the algorithm as a whole). That is also the reason that we always run each algorithm for the same number of iterations (i.e., at most 10,000).

Validity: Conclusions drawn from the experimental results are based on the correct interpretations of data.

During the analysis of the results we were careful to find the valid cause (often confirmed with a theoretical explanation) for the achieved speedups. This is mainly reflected in the engineering process described in Section 4, where in each phase we dealt with one issue causing a performance bottleneck. Moreover, experiments with the tableau density establish a perspective on the practical applicability of the proposed data-flow algorithms.

Generality: Analysis of the results and conclusions should apply broadly. The experiment should be carefully designed by choosing proper performance measures and enable a deduction of useful results.

In our experiments we thrived to isolate various data-flow parameter and techniques in order to deduce which ones have a positive effect on the performance. For example, the above experiments clearly show the positive effect of pipes and use of large memory.

Efficiency: To produce correct results without wasting time and other resources, i.e., maximize information gained per unit of experimental effort.

Data-flow programs are known to have long compile times. Hence, before performing the workhorse experiments, we extensively used the Maxeler's data-flow engine simulator to check correctness and obtain preliminary results. Moreover, our test program is flexible enough (via support of various command line parameters) to support testing in various experimental settings.

Newsworthiness: Finally, we strongly believe that we produced interesting results and conclusions. First, we showed that the simplex algorithm can be successfully transferred to the data-flow architecture. Next, via the control of parameters we explored various data-flow programming techniques, and, finally, we showed the practical applicability of our approach.

6. CONCLUSIONS AND FUTURE WORK

In this chapter we presented an implementation of the primal simplex algorithm for the data-flow architecture. The algorithm is very suitable for this platform and the speedups are significant compared to the control-flow-based implementation. The empirical results presented in Section 5 show the algorithm to be even too data-intensive, resulting in data transfer bottlenecks, which ultimately limit the amount of parallelism that can be achieved.

Even though the revised simplex method is not as suitable for the dataflow paradigm as the primal simplex algorithm, there are various possibilities for implementing it. In the future work we will investigate the possibilities of adapting it and compare with the state-of-the-art solvers. We will also investigate the extensions to the primal algorithm on platforms with several DFEs, where a more global data orchestration is required to achieve even higher speedups.

ACKNOWLEDGMENTS

This work was done during our visit to the Mathematical Institute SANU and the Electrotechnical Faculty of the University of Belgrade. We would like to thank prof. Veljko Milutinović and his colleagues for the hospitality, for providing us with the required infrastructure and giving us many insightful guidelines for our work. We would also like to thank the reviewers for constructive comments and suggestions for improving this work.

REFERENCES

[1] A.R. Hurson, V. Milutinović, Dataflow Processing, Advances in Computers, vol. 96, Elsevier, The Netherlands, 2015, pp. vii–viii.
[2] Maxeler Technologies, Maximum Performance Computing. http://www.maxeler.com (accessed 17.02.17).
[3] A. Kos, S. Tomažič, J. Salom, N. Trifunovic, M. Valero, V. Milutinovic, New benchmarking methodology and programming model for big data processing, Int. J. Distrib. Sen. Netw. 2015 (2015) 4:4, ISSN: 1550-1329, http://dx.doi.org/10.1155/2015/271752.
[4] M.J. Flynn, O. Mencer, V. Milutinović, G. Rakočević, P. Stenstrom, R. Trobec, M. Valero, Moving from petaflops to petadata, Commun. ACM 56 (5) (2013) 39–42.
[5] N. Trifunovic, V. Milutinovic, J. Salom, A. Kos, Paradigm shift in big data supercomputing: dataflow vs. controlflow, J. Big Data 2 (1) (2015) 1–9.
[6] A. Kos, V. Ranković, S. Tomažič, Sorting networks on Maxeler Dataflow supercomputing systems, Adv. Comput. 96 (2015) 139–186.
[7] V. Ranković, A. Kos, V. Milutinović, Bitonic merge sort implementation on the Maxeler Dataflow supercomputing system, IPSI BgD Trans. Internet Res. 9 (2) (2013) 5–10.
[8] V. Milutinović, J. Salom, N. Trifunović, R. Giorgi, An example application: Fourier transform, in: Guide to DataFlow Supercomputing, Springer, Cham, Switzerland, 2015, pp. 73–106.
[9] D. Oriato, S. Tilbury, M. Marrocu, G. Pusceddu, Acceleration of a meteorological limited area model with dataflow engines, in: Symposium on Application Accelerators in High Performance Computing (SAAHPC), 2012, IEEE, 2012, pp. 129–132.
[10] B.A. Cipra, The best of the 20th century: editors name top 10 algorithms, SIAM News 33 (2000) 1–2.
[11] J. Eckstein, İ. Boduroğlu, L.C. Polymenakos, D. Goldfarb, Data-parallel implementations of dense simplex methods on the connection machine CM-2, ORSA J. Comput. 7 (4) (1995) 402–416, http://dx.doi.org/10.1287/ijoc.7.4.402.

[12] G. Yarmish, The simplex method applied to wavelet decomposition, in: Proceedings of the 10th WSEAS International Conference on Applied Mathematics, Dallas, Texas, MATH'06, World Scientific and Engineering Academy and Society (WSEAS), Stevens Point, WI, ISBN 999-6666-22-12006, pp. 226–228, http://dl.acm.org/citation.cfm?id=1376439.1376477.

[13] S.S. Morgan, A Comparison of Simplex Method Algorithms, University of Florida, Gainesville, FL, 1997, https://books.google.rs/books?id=FwxuJQAACAAJ.

[14] V. Milutinović, J. Salom, N. Trifunovic, R. Giorgi, Guide to DataFlow Supercomputing: Basic Concepts, Case Studies, and a Detailed Example, Springer International Publishing, Cham, 2015.

[15] K.L. Hoffman, M. Padberg, Solving airline crew scheduling problems by branch-and-cut, Manag. Sci. 39 (6) (1993) 657–682.

[16] S.C. Myers, A note on linear programming and capital budgeting, J. Financ. 27 (1) (1972) 89–92.

[17] T.S. Hale, C.R. Moberg, Location science research: a review, Ann. Oper. Res. 123 (1–4) (2003) 21–35.

[18] S. Tayur, R. Ganeshan, M. Magazine, Quantitative Models for Supply Chain Management, International Series in Operations Research & Management Science, vol. 17, Springer Science & Business Media, Dordrecht, The Netherlands, 2012.

[19] T.H. Cormen, C. Stein, R.L. Rivest, C.E. Leiserson, Introduction to Algorithms, second ed., McGraw-Hill Higher Education, New York, NY, 2001, ISBN 0070131511.

[20] S. Boyd, L. Vandenberghe, Convex Optimization, Cambridge University Press, Cambridge, 2004.

[21] T. Terlaky, S. Zhang, Pivot rules for linear programming: a survey on recent theoretical developments, Ann. Oper. Res. 46–47 (1) (1993) 203–233, ISSN: 0254-5330, http://dx.doi.org/10.1007/BF02096264.

[22] D.A. Bader, B.M.E. Moret, P. Sanders, Algorithm engineering for parallel computation, in: R. Fleischer, B. Moret, E.M. Schmidt (Eds.), Experimental Algorithmics, Springer-Verlag New York, Inc., New York, NY, ISBN 3-540-00346-0, 2002, pp. 1–23, http://dl.acm.org/citation.cfm?id=857152.857154.

[23] C.C. McGeoch, A Guide to Experimental Algorithmics, first ed., Cambridge University Press, New York, NY, 2012, ISBN 0521173019, 9780521173018.

[24] M. Müller-Hannemann, S. Schirra, Algorithm Engineering: Bridging the Gap Between Algorithm Theory and Practice, Lecture Notes in Computer Science, Springer, Berlin, Heidelberg, 2010, ISBN 9783642148668, https://books.google.si/books?id=cttsCQAAQBAJ.

[25] PCI-SIG: Peripheral Component Interconnect Special Interest Group. http://pcisig.com/, 2015 (accessed 17.02.17).

[26] M. Berkelaar, K. Eikland, P. Notebaert, Open Source (Mixed-Integer) Linear Programming System. http://lpsolve.sourceforge.net/, 2004 (accessed 17.02.17).

[27] J. Mihelič, U. Čibej, Experimental algorithmics for the dataflow architecture: guidelines and issues, IPSI BgD Trans. Adv. Res. 13 (1) (2017) 1–8.

[28] J. Mihelič, U. Čibej, Simplex Algorithm for the Maxeler Data-Flow Computer Architecture. https://github.com/jurem/DataFlowSimplex (accessed 17.02.17).

[29] Maxeler Technologies, Maxeler AppGallery. http://appgallery.maxeler.com/ (accessed 17.02.17).

ABOUT THE AUTHORS

Uroš Čibej received his doctoral degree in computer science from the University of Ljubljana in 2007. Currently he is with the Laboratory of Algorithms and Data structures. His research interests include location problems, distributed systems, computational models, halting probability, graph algorithms, and computational complexity.

Jurij Mihelič received the doctoral degree in computer science from the University of Ljubljana, Slovenia, in 2006. Currently, he is with the Laboratory of Algorithms and Data Structures, Faculty of Computer and Information Science, University of Ljubljana, Slovenia, as an assistant professor and a researcher. His research interests include combinatorial optimization, heuristics, approximation algorithms, graph problems, and uncertainty in optimization problems as well as algorithm engineering and experimental algorithmics.

CHAPTER FOUR

Simple Operations in Memory to Reduce Data Movement

Vivek Seshadri*, Onur Mutlu[†]
*Microsoft Research India, Bangalore, India
[†]ETH Zürich, Zürich, Switzerland

Contents

Advances in Computers, Volume 106
ISSN 0065-2458
http://dx.doi.org/10.1016/bs.adcom.2017.04.004

Abstract

In existing systems, the off-chip memory interface allows the memory controller to perform only read or write operations. Therefore, to perform any operation, the processor must first read the source data and then write the result back to memory after performing the operation. This approach consumes high latency, bandwidth, and energy for operations that work on a large amount of data. Several works have proposed techniques to process data near memory by adding a small amount of compute logic closer to the main memory chips. In this chapter, we describe two techniques proposed by recent works that take this approach of processing in memory further by exploiting the underlying operation of the main memory technology to perform more complex tasks. First, we describe RowClone, a mechanism that exploits DRAM technology to perform bulk copy and initialization operations completely inside main memory. We then describe a complementary work that uses DRAM to perform bulk bitwise AND and OR operations inside main memory. These two techniques significantly improve the performance and energy efficiency of the respective operations.

ABBREVIATIONS

DDR double data rate
DMA direct memory access
DRAM dynamic random-access memory
EEPROM electrically erasable programmable read-only memory
PCM phase-change memory
PiM processing in memory
PuM processing using memory

1. INTRODUCTION

In modern systems, the channel that connects the processor and off-chip main memory is a critical bottleneck for both performance and energy efficiency. First, the channel has limited data bandwidth. Increasing this available bandwidth requires increasing the number of channels or the width of each channel or the frequency of the channel. All these approaches significantly increase the cost of the system and are not scalable. Second, a significant fraction of the energy consumed in performing an operation is spent on moving data over the off-chip memory channel [1].

To address this problem, many prior and recent works [2–28] have proposed techniques to process data near memory, an approach widely referred to as *processing in memory* or PiM. The idea behind PiM is to add a small amount of compute logic close to the memory chips and use that logic to perform simple yet bandwidth-intensive and/or latency-sensitive operations. The premise is that being close to the memory chips, the PiM module will have much higher bandwidth and lower latency to memory than the regular processor. Consequently, PiM can (1) perform bandwidth-intensive and latency-sensitive operations faster and (2) reduce the off-chip memory bandwidth requirements of such operations. As a result, PiM significantly improves both overall system performance and energy efficiency.

In this chapter, we focus our attention on two works that push the notion of PiM deeper by exploiting the underlying operation of the main memory technology to perform more complex tasks. We will refer to this approach as *Processing using Memory* or PuM. Unlike PiM, which adds new logic structures near memory to perform computation, the key idea behind PuM is to exploit some of the peripheral structures *already* existing inside memory devices (with minimal changes) to perform other tasks.

The first work that we will discuss in this chapter is RowClone [25], a mechanism that exploits dynamic random-access memory (DRAM) technology to perform bulk data copy and initialization completely inside DRAM. Such bulk copy and initialization operations are triggered by many applications (e.g., bulk zeroing) and system-level functions (e.g., page copy operations). Despite the fact that these operations require no computation, existing system must necessarily read and write the required data over the main memory channel. In fact, even with a high-speed memory bus (DDR4-2133) a simple 4 KB copy operation can take close to half a micro second for just the data transfers on the memory channel. By performing such operations completely inside main memory, RowClone eliminates the need for any data transfer on the memory channel, thereby significantly improving performance and energy efficiency.

The second work that we will discuss in this chapter is a mechanism to perform bulk bitwise AND and OR operations completely inside DRAM [24]. Bitwise operations are an important component of modern day programming. Many applications (e.g., bitmap indices) rely on bitwise operations on large bitvectors to achieve high performance. Similar to bulk copy or initialization, the throughput of bulk bitwise operations in existing systems is also limited by the available memory bandwidth. The *In-DRAM AND-OR* mechanism (IDAO) avoids the need to transfer large amounts of data on the memory channel to perform these operations. Similar to RowClone, IDAO enables an order of magnitude improvement in the performance of bulk bitwise operations. We will describe these two works in detail in this chapter.

In this chapter, we will discuss the following things.

- We motivate the need for reducing data movement and how processing near memory helps in achieving that goal (Section 2). We will briefly describe a set of recent works that have pushed the idea of processing near memory deeper by using the underlying memory technologies (e.g., DRAM, spin-torque-transfer magnetic random-access memory, and phase-change memory [PCM]) to perform tasks more complex than just storing data (Section 3).

- As the major focus of this chapter is on the PuM works that build on DRAM, we provide a brief background on modern DRAM organization and operation that is sufficient to understand the mechanisms (Section 4).

- We describe the two mechanisms, RowClone (in-DRAM bulk copy and initialization) and In-DRAM-AND-OR (in-DRAM bulk bitwise AND and OR) in detail in Sections 5 and 6, respectively.

- We describe a number of applications for the two mechanisms and quantitative evaluations showing that they improve performance and energy efficiency compared to existing systems.

2. PROCESSING IN MEMORY

Data movement contributes a major fraction of the execution time and energy consumption of many programs. The farther the data is from the processing engine (e.g., CPU), the more the contribution of data movement toward execution time and energy consumption. While most programs aim to keep their active working set as close to the processing engine as possible (say the L1 cache), for applications with working sets larger than the on-chip cache size, the data typically reside in main memory.

Unfortunately, main memory latency is not scaling commensurately with the remaining resources in the system, namely, the compute power and memory capacity. As a result, the performance of most large-working-set applications is limited by main memory latency and/or bandwidth. For instance, just transferring a single page (4 KB) of data from DRAM can consume between a quarter and half a microsecond even with high speed memory interfaces (DDR4-2133 [29]). During this time, the processor can potentially execute hundreds to thousands of instructions. With respect to energy, while performing a 64-bit double precision floating point operation typically consumes few tens of pico-joules, accessing 64 bits of data from off-chip DRAM consumes few tens of nano-joules (three orders of magnitude more energy) [1].

One of the solutions to address this problem is to add support to process data closer to memory, especially for operations that access large amounts of data. This approach is generally referred to as *PiM* or *near data processing*. The high-level idea behind PiM is to add a small piece of compute logic closer to memory that has much higher bandwidth to memory than the main processor. Prior research has proposed two broad ways of implementing PiM: (1) Integrating processing logic into the memory chips, and (2) using 3D-stacked memory architectures.

2.1 Integrating Processing Logic in Memory

Many works (e.g., Logic-in-Memory Computer [17], NON-VON Database Machine [18], EXECUBE [20], Terasys [22], Intelligent RAM [19], Active Pages [21], FlexRAM [30, 31], Computational RAM [23], and DIVA [32]) have proposed mechanisms and models to add processing logic close to memory. The idea is to integrate memory and CPU on the same chip by designing the CPU using the memory process technology. The reduced data movement allows these approaches to enable low-latency, high-bandwidth, and low-energy data communication. However, they suffer from two key shortcomings.

First, this approach of integrating processor on the same chip as memory greatly increases the overall cost of the system. Second, DRAM vendors use a high-density process to minimize cost-per-bit. Unfortunately, high-density DRAM process is not suitable for building high-speed logic [19]. As a result, this approach is not suitable for building a general purpose processor near memory, at least with modern logic and high-density DRAM technologies.

2.2 3D-Stacked DRAM Architectures

Some recent DRAM architectures [16, 33–35] use 3D-stacking technology to stack multiple DRAM chips on top of the processor chip or a separate logic layer. These architectures offer much higher bandwidth to the logic layer compared to traditional off-chip interfaces. This enables an opportunity to offload some computation to the logic layer, thereby improving performance. In fact, many recent works have proposed mechanisms to improve and exploit such architectures (e.g., [2–16, 26, 36, 37]). 3D-stacking enables much higher bandwidth between the logic layer and the memory chips, compared to traditional architectures. However, 3D-stacked architectures still require data to be transferred outside the DRAM chip, and hence can be bandwidth-limited. In addition, thermal factors constrain the number of chips that can be stacked, thereby limiting the memory capacity. As a result, multiple 3D-stacked DRAMs are required to scale to large workloads. Despite these limitations, this approach seems to be the most viable way of implementing PiM in modern systems.

3. PROCESSING USING MEMORY

In this chapter, we introduce a new class of work that pushes the idea of PiM further by exploiting the underlying memory operation to perform more complex operations than just data storage. We refer to this class of works as *processing using memory* (*PuM*).

Reducing cost-per-bit is a first-order design constraint for most memory technologies. As a result, the memory cells are small. Therefore, most memory devices use significant amount of sensing and peripheral logic to extract data from the memory cells. The key idea behind PuM is to use these logic structures and their operation to perform some additional tasks.

It is clear that unlike PiM, which can potentially be designed to perform any task, PuM can only enable some limited functionality. However, for tasks that can be performed by PuM, PuM has two advantages over PiM. First, as PuM exploits the underlying operation of memory, it incurs much lower cost than PiM. Second, unlike PiM, PuM does not have to read any data out of the memory chips. As a result, the PuM approach is possibly the most energy efficient way of performing the respective operations.

Building on top of DRAM, which is the technology ubiquitously used to build main memory in modern systems, two recent works take the PuM approach to accelerate certain important primitives: (1) RowClone [25], which performs bulk copy and initialization operations completely inside DRAM, and (2) IDAO [24], which performs bulk bitwise AND/OR operations completely inside DRAM. Both these works exploit the operation of the DRAM sense amplifier and the internal organization of DRAM to perform the respective operations. We will discuss these two works in significant detail in this chapter.

Similar to these works, there are others that build on various other memory technologies. Pinatubo [38] exploits PCM [39–45] architecture to perform bitwise operations efficiently inside PCM. Pinatubo enhances the PCM sense amplifiers to sense fine grained differences in resistance and use this to perform bitwise operations on multiple cells connected to the same sense amplifier. As we will describe in this chapter, bitwise operations are critical for many important data structures like bitmap indices. Kang et al. [46] propose a mechanism to exploit static random-access memory architecture to accelerate the primitive "sum of absolute differences". ISAAC [47] is a mechanism to accelerate vector dot product operations using a memristor array. ISAAC uses the crossbar structure of a memristor array and its analog operation to efficiently perform dot products. These operations are heavily used in many important applications including deep neural networks.

In the subsequent sections, we will focus our attention on RowClone and IDAO. We will first provide the necessary background on DRAM design and then describe how these mechanisms work.

4. BACKGROUND ON DRAM

In this section, we describe the necessary background to modern DRAM architecture and its implementation. While we focus our attention primarily on commodity DRAM design (i.e., the DDRx interface), most DRAM architectures use very similar design approaches and vary only in higher-level design choices. As a result, the mechanisms we describe in the subsequent sections can be extended to any DRAM architecture. There has been significant recent research in DRAM architectures and the interested reader can find details about various aspects of DRAM in multiple recent publications [48–60].

4.1 High-Level Organization of the Memory System

Fig. 1 shows the organization of the memory subsystem in a modern system. At a high level, each processor chip consists of one of more off-chip memory *channels*. Each memory channel consists of its own set of *command*, *address*, and *data* buses. Depending on the design of the processor, there can be either

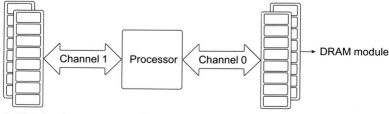

Fig. 1 High-level organization of the memory subsystem.

an independent memory controller for each memory channel or a single memory controller for all memory channels. All modules connected to a channel share the buses of the channel. Each module consists of many DRAM devices (or chips). Most of this section is dedicated to describing the design of a modern DRAM chip. In Section 4.3, we present more details of the module organization of commodity DRAM.

4.2 DRAM Chip

A modern DRAM chip consists of a hierarchy of structures: DRAM *cells*, *tiles/MATs*, *subarrays*, and *banks*. In this section, we describe the design of a modern DRAM chip in a bottom-up fashion, starting from a single DRAM cell and its operation.

4.2.1 DRAM Cell and Sense Amplifier

At the lowest level, DRAM technology uses capacitors to store information. Specifically, it uses the two extreme states of a capacitor, namely, the *empty* and the *fully charged* states to store a single bit of information. For instance, an empty capacitor can denote a logical value of 0, and a fully charged capacitor can denote a logical value of 1. Fig. 2 shows the two extreme states of a capacitor.

Unfortunately, the capacitors used for DRAM chips are small and will get smaller with each new generation. As a result, the amount of charge that can be stored in the capacitor, and hence the difference between the two states is also very small. In addition, the capacitor can potentially lose its state after it is accessed. Therefore, to extract the state of the capacitor, DRAM manufacturers use a component called *sense amplifier*.

Fig. 3 shows a sense amplifier. A sense amplifier contains two inverters which are connected together such that the output of one inverter is connected to the input of the other and vice versa. The sense amplifier also has an enable signal that determines if the inverters are active. When enabled, the sense amplifier has two stable states, as shown in Fig. 4. In both these

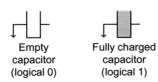

Empty
capacitor
(logical 0)

Fully charged
capacitor
(logical 1)

Fig. 2 Two states of a DRAM cell.

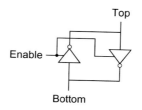

Fig. 3 Sense amplifier.

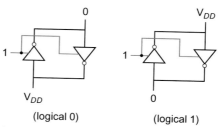

(logical 0)   (logical 1)

Fig. 4 Stable states of a sense amplifier.

stable states, each inverter takes a logical value and feeds the other inverter with the negated input.

Fig. 5 shows the operation of the sense amplifier from a disabled state. In the initial disabled state, we assume that the voltage level of the top terminal (V_a) is higher than that of the bottom terminal (V_b). When the sense amplifier is enabled in this state, it *senses* the difference between the two terminals and *amplifies* the difference until it reaches one of the stable states (hence the name "sense amplifier").

4.2.2 DRAM Cell Operation: The Activate-Precharge Cycle

DRAM technology uses a simple mechanism that converts the logical state of a capacitor into a logical state of the sense amplifier. Data can then be accessed from the sense amplifier (since it is in a stable state). Fig. 6 shows the connection between a DRAM cell and the sense amplifier and the sequence of states involved in converting the cell state into the sense amplifier state.

As shown in the figure (state ❶), the capacitor is connected to an access transistor that acts as a switch between the capacitor and the sense amplifier. The transistor is controlled by a wire called *wordline*. The wire that connects the transistor to the top end of the sense amplifier is called *bitline*. In the initial state ❶, the wordline is lowered, the sense amplifier is disabled and both ends

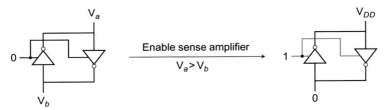

Fig. 5 Operation of the sense amplifier.

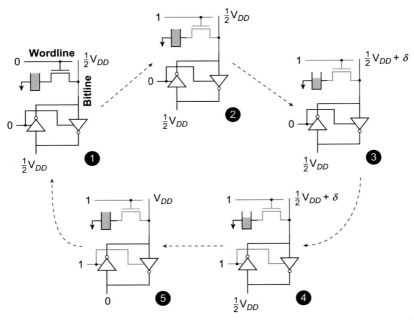

Fig. 6 Operation of a DRAM cell and sense amplifier.

of the sense amplifier are maintained at a voltage level of $\frac{1}{2}V_{DD}$. We assume that the capacitor is initially fully charged (the operation is similar if the capacitor was empty). This state is referred to as the *precharged* state. An access to the cell is triggered by a command called ACTIVATE. Upon receiving an ACTIVATE, the corresponding wordline is first raised (state ❷). This connects the capacitor to the bitline. In the ensuing phase called *charge sharing* (state ❸), charge flows from the capacitor to the bitline, raising the voltage level on the bitline (top end of the sense amplifier) to $\frac{1}{2}V_{DD} + \delta$. After charge sharing, the sense amplifier is enabled (state ❹). The sense amplifier detects the difference in voltage levels between its two ends and amplifies the deviation, till it reaches the stable state where the top end is at V_{DD} (state ❺). Since the capacitor is still connected to the bitline, the charge on the capacitor is also fully restored. We shortly describe how the data can be accessed from the sense amplifier. However, once the access to the cell is complete, the cell is taken back to the original precharged state using the command called PRECHARGE. Upon receiving a PRECHARGE, the wordline is first lowered, thereby disconnecting the cell from the sense amplifier. Then, the two ends of the sense amplifier are driven to $\frac{1}{2}V_{DD}$ using a precharge unit (not shown in the figure for brevity).

4.2.3 DRAM MAT/Tile: The Open Bitline Architecture

A major goal of DRAM manufacturers is to maximize the density of the DRAM chips while adhering to certain latency constraints (described in Section 4.2.6). There are two costly components in the setup described in the previous section. The first component is the sense amplifier itself. Each sense amplifier is around two orders of magnitude larger than a single DRAM cell [49, 61]. Second, the state of the wordline is a function of the address that is currently being accessed. The logic that is necessary to implement this function (for each cell) is expensive.

In order to reduce the overall cost of these two components, they are shared by many DRAM cells. Specifically, each sense amplifier is shared by a column of DRAM cells. In other words, all the cells in a single column are connected to the same bitline. Similarly, each wordline is shared by a row of DRAM cells. Together, this organization consists of a 2D array of DRAM cells connected to a row of sense amplifiers and a column of wordline drivers. Fig. 7 shows this organization with a 4 × 4 2D array.

To further reduce the overall cost of the sense amplifiers and the wordline driver, modern DRAM chips use an architecture called the *open bitline architecture*. This architecture exploits two observations. First, the sense amplifier is wider than the DRAM cells. This difference in width results in a white space near each column of cells. Second, the sense amplifier is symmetric. Therefore, cells can also be connected to the bottom part of the sense amplifier. Putting together these two observations, we can pack twice as many cells in the same area using the open bitline architecture, as shown in Fig. 8.

As shown in the figure, a 2D array of DRAM cells is connected to two rows of sense amplifiers: one on the top and one on the bottom of the array.

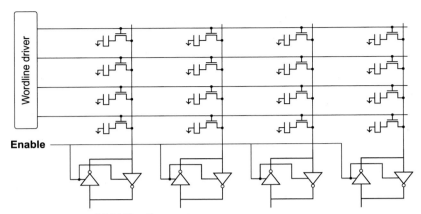

Fig. 7 A 2D array of DRAM cells.

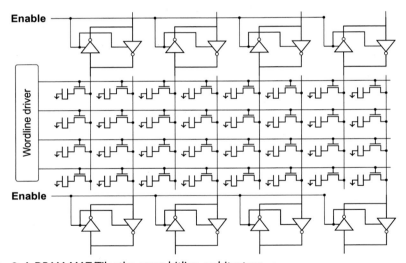

Fig. 8 A DRAM MAT/Tile: the open bitline architecture.

While all the cells in a given row share a common wordline, half the cells in each row are connected to the top row of sense amplifiers and the remaining half of the cells are connected to the bottom row of sense amplifiers. This tightly packed structure is called a DRAM MAT/Tile [48, 62, 63]. In a modern DRAM chip, each MAT typically is a 512 × 512 or 1024 × 1024 array. Multiple MATs are grouped together to form a larger structure called a *DRAM bank*, which we describe next.

4.2.4 DRAM Bank

In most modern commodity DRAM interfaces [29, 64], a DRAM bank is the smallest structure visible to the memory controller. All commands related to data access are directed to a specific bank. Logically, each DRAM bank is a large monolithic structure with a 2D array of DRAM cells connected to a single set of sense amplifiers (also referred to as a row buffer). For example, in a 2 Gb DRAM chip with 8 banks, each bank has 2^{15} rows and each logical row has 8192 DRAM cells. Fig. 9 shows this logical view of a bank.

In addition to the MAT, the array of sense amplifiers, and the wordline driver, each bank also consists of some peripheral structures to decode DRAM commands and addresses, and manage the input/output to the DRAM bank. Specifically, each bank has a *row decoder* to decode the row address of row-level commands (e.g., ACTIVATE). Each data access command (READ and WRITE) accesses only a part of a DRAM row. Such individual parts are referred to as *columns*. With each data access command, the address of the column to be accessed is provided. This address is decoded by the *column selection logic*. Depending on which column is selected, the corresponding piece of data is communicated between the sense amplifiers and the bank

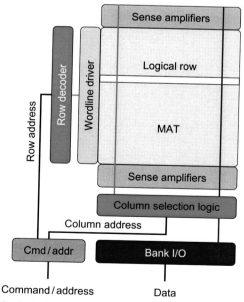

Fig. 9 DRAM bank: logical view.

I/O logic. The bank I/O logic in turn acts as an interface between the DRAM bank and the chip-level I/O logic.

Although the bank can logically be viewed as a single MAT, building a single MAT of a very large dimension is practically not feasible as it will require very long bitlines and wordlines. Therefore, each bank is physically implemented as a 2D array of DRAM MATs. Fig. 10 shows a physical implementation of the DRAM bank with 4 MATs arranged in 2 × 2 array. As shown in the figure, the output of the global row decoder is sent to each row of MATs. The bank I/O logic, also known as the *global sense amplifiers*, are connected to all the MATs through a set of *global bitlines*. As shown in the figure, each vertical collection of MATs consists of its own columns selection logic and global bitlines. In a real DRAM chip, the global bitlines run on top of the MATs in a separate metal layer. One implication of this division is that the data accessed by any command is split equally across all the MATs in a single row of MATs.

Fig. 11 shows the zoomed-in version of a DRAM MAT with the surrounding peripheral logic. Specifically, the figure shows how each column selection line selects specific sense amplifiers from an MAT and connects

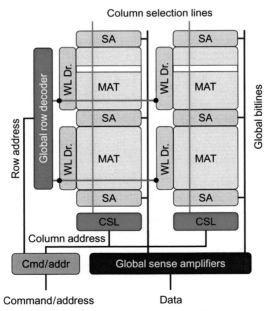

Fig. 10 DRAM bank: physical implementation. In a real chip, the global bitlines run on top of the MATs in a separate metal layer (components in figure are not to scale).

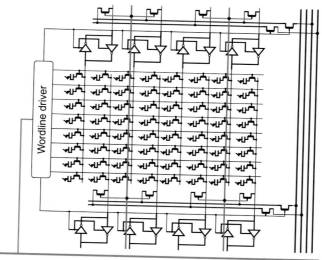

Fig. 11 Detailed view of an MAT.

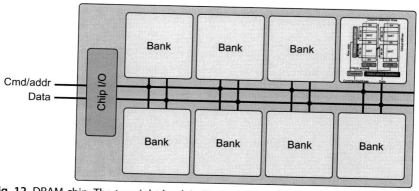

Fig. 12 DRAM chip. The top-right bank in Figure 12 essentially contains a zoomed-out version of Figure 10.

them to the global bitlines. It should be noted that the width of the global bitlines for each MAT (typically 8/16) is much smaller than that of the width of the MAT (typically 512/1024). This is because the global bitlines span a much longer distance and hence have to be thicker to ensure integrity.

Each DRAM chip consist of multiple banks as shown in Fig. 12. All the banks share the chip's internal command, address, and data buses. As mentioned before, each bank operates mostly independently (except for operations that involve the shared buses). The chip I/O manages the transfer of data to and from the chip's internal bus to the memory channel. The width of the chip output (typically 8 bits) is much smaller than the output width of each bank (typically 64 bits). Any piece of data accessed from a DRAM bank is first buffered at the chip I/O and sent out on the memory bus 8 bits at a time. With the DDR (double data rate) technology, 8 bits are sent out each half cycle. Therefore, it takes four cycles to transfer 64 bits of data from a DRAM chip I/O on to the memory channel.

4.2.5 DRAM Commands: Accessing Data From a DRAM Chip

To access a piece of data from a DRAM chip, the memory controller must first identify the location of the data: the bank ID (B), the row address (R) within the bank, and the column address (C) within the row. After identifying these pieces of information, accessing the data involves three steps.

The first step is to issue a PRECHARGE to the bank B. This step prepares the bank for a data access by ensuring that all the sense amplifiers are in the *precharged* state (Fig. 6, state ❶). No wordline within the bank is raised in this state.

The second step is to activate the row R that contains the data. This step is triggered by issuing a ACTIVATE to bank B with row address R. Upon receiving this command, the corresponding bank feeds its global row decoder with the input R. The global row decoder logic then raises the wordline of the DRAM row corresponding to the address R and enables the sense amplifiers connected to that row. This triggers the DRAM cell operation described in Section 4.2.2. At the end of the activate operation the data from the entire row of DRAM cells are copied to the corresponding array of sense amplifiers.

Finally, the third step is to access the data from the required column. This is done by issuing a READ or WRITE command to the bank with the column address C. Upon receiving a READ or WRITE command, the corresponding address is fed to the column selection logic. The column selection logic then raises the column selection lines (Fig. 11) corresponding to address C, thereby connecting those sense amplifiers to the global sense amplifiers through the global bitlines. For a read access, the global sense amplifiers sense the data from the MAT's local sense amplifiers and transfer that data to the chip's internal bus. For a write access, the global sense amplifiers read the data from the chip's internal bus and force the MAT's local sense amplifiers to the appropriate state.

Not all data accesses require all three steps. Specifically, if the row to be accessed is already activated in the corresponding bank, then the first two steps can be skipped and the data can be directly accessed by issuing a READ or WRITE to the bank. For this reason, the array of sense amplifiers is also

Table 1 Key DRAM Timing Constraints With Their Values for DDR3-1600

Name	Constraint		Description	Value (ns)
tRAS	ACTIVATE	→ PRECHARGE	Time taken to complete a row activation operation in a bank	35
tRCD	ACTIVATE	→ READ/ WRITE	Time between an activate command and column command to a bank	15
tRP	PRECHARGE	→ ACTIVATE	Time taken to complete a precharge operation in a bank	15
tWR	WRITE	→ PRECHARGE	Time taken to ensure that data is safely written to the DRAM cells after a write operation (*write recovery*)	15

referred to as a *row buffer*, and such an access that skips the first two steps is called a *row buffer hit*. Similarly, if the bank is already in the precharged state, then the first step can be skipped. Such an access is referred to as a *row buffer miss*. Finally, if a different row is activated within the bank, then all three steps have to be performed. Such an access is referred to as a *row buffer conflict*.

4.2.6 DRAM Timing Constraints

Different operations within DRAM consume different amounts of time. Therefore, after issuing a command, the memory controller must wait for a sufficient amount of time before it can issue the next command. Such wait times are managed by what are called the *timing constraints*. Timing constraints essentially dictate the minimum amount of time between two commands issued to the same bank/rank/channel. Table 1 describes some key timing constraints along with their values for the DDR3-1600 interface.

4.3 DRAM Module

As mentioned before, each READ or WRITE command for a single DRAM chip typically involves only 64 bits. In order to achieve high memory bandwidth, commodity DRAM modules group several DRAM chips (typically 4 or 8) together to form a *rank* of DRAM chips. The idea is to connect all chips of a single rank to the same command and address buses, while providing each chip with an independent data bus. In effect, all the chips within a rank receive the same commands with same addresses, making the rank a logically wide DRAM chip.

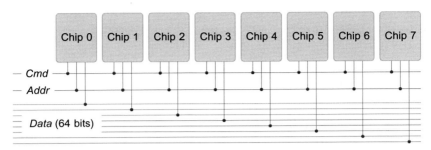

Fig. 13 Organization of a DRAM rank.

Fig. 13 shows the logical organization of a DRAM rank. Most commodity DRAM ranks consist of eight chips. Therefore, each READ or WRITE command accesses 64 bytes of data, the typical cache line size in most processors.

5. ROWCLONE

In this section, we present RowClone [25], a mechanism that can perform bulk copy and initialization operations completely inside DRAM. This approach obviates the need to transfer large quantities of data on the memory channel, thereby significantly improving the efficiency of a bulk copy operation. As bulk data initialization (specifically bulk zeroing) can be viewed as a special case of a bulk copy operation, RowClone can be easily extended to perform such bulk initialization operations with high efficiency.

RowClone consists of two independent mechanisms that exploit several observations about DRAM organization and operation. The first mechanism, called the *Fast Parallel Mode* (FPM), efficiently copies data between two rows of DRAM cells that share the same set of sense amplifiers (i.e., two rows within the same subarray). The second mechanism, called the *Pipelined Serial Mode*, efficiently copies cache lines between two banks within a module in a pipelined manner. Although not as fast as FPM, PSM has fewer constraints and hence is more generally applicable. We now describe these two mechanisms in detail.

5.1 Fast Parallel Mode

The FPM is based on the following three observations about DRAM.

1. In a commodity DRAM module, each ACTIVATE command transfers data from a large number of DRAM cells (multiple kilobytes) to the corresponding array of sense amplifiers (Section 4.3).

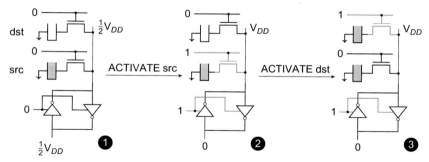

Fig. 14 RowClone: fast parallel mode.

2. Several rows of DRAM cells share the same set of sense amplifiers (Section 4.2.3).

3. A DRAM cell is not strong enough to flip the state of the sense amplifier from one stable state to another stable state. In other words, if a cell is connected to an already activated sense amplifier (or bitline), then the data of the cell gets overwritten with the data on the sense amplifier.

While the first two observations are direct implications from the design of commodity DRAM, the third observation exploits the fact that DRAM cells can cause only a small perturbation on the bitline voltage. Fig. 14 pictorially shows how this observation can be used to copy data between two cells that share a sense amplifier.

The figure shows two cells (src and dst) connected to a single sense amplifier. In the initial state, we assume that src is fully charged and dst is fully empty, and the sense amplifier is in the precharged state (❶). In this state, FPM issues an ACTIVATE to src. At the end of the activation operation, the sense amplifier moves to a stable state where the bitline is at a voltage level of V_{DD} and the charge in src is fully restored (❷). FPM follows this operation with an ACTIVATE to dst, without an intervening PRECHARGE. This operation lowers the wordline corresponding to src and raises the wordline of dst, connecting dst to the bitline. Since the bitline is already fully activated, even though dst is initially empty, the perturbation caused by the cell is not sufficient to flip the state of the bitline. As a result, the sense amplifier continues to drive the bitline to V_{DD}, thereby pushing dst to a fully charged state (❸).

It can be shown that regardless of the initial state of src and dst, the above operation copies the data from src to dst. Given that each ACTIVATE operates on an entire row of DRAM cells, the above operation can copy multiple kilobytes of data with just two back-to-back ACTIVATE operations.

Unfortunately, modern DRAM chips do not allow another ACTIVATE to an already activated bank—the expected result of such an action is undefined. This is because a modern DRAM chip allows at most one row (subarray) within each bank to be activated. If a bank that already has a row (subarray) activated receives an ACTIVATE to a different subarray, the currently activated subarray must first be precharged [48]. Some DRAM manufacturers design their chips to drop back-to-back ACTIVATEs to the same bank.

To support FPM, RowClone changes the way a DRAM chip handles back-to-back ACTIVATEs to the same bank. When an already activated bank receives an ACTIVATE to a row, the chip allows the command to proceed if and only if the command is to a row that belongs to the currently activated subarray. If the row does not belong to the currently activated subarray, then the chip takes the action it normally does with back-to-back ACTIVATEs—e.g., drop it. Since the logic to determine the subarray corresponding to a row address is already present in today's chips, implementing FPM only requires a comparison to check if the row address of an ACTIVATE belongs to the currently activated subarray, the cost of which is almost negligible.

Summary. To copy data from src to dst within the same subarray, FPM first issues an ACTIVATE to src. This copies the data from src to the subarray row buffer. FPM then issues an ACTIVATE to dst. This modifies the input to the subarray row decoder from src to dst and connects the cells of dst row to the row buffer. This, in effect, copies the data from the sense amplifiers to the destination row. With these two steps, FPM can copy a 4 KB page of data $12.0\times$ faster and with $74.4\times$ less energy than an existing system (we describe the methodology in Section 8.1).

Limitations. FPM has two constraints that limit its general applicability. First, it requires the source and destination rows to be within the same subarray (i.e., share the same set of sense amplifiers). Second, it cannot partially copy data from one row to another. Despite these limitations, FPM can be immediately applied to today's systems to accelerate two commonly used primitives in modern systems—copy-on-write and bulk zeroing. In the following section, we describe the second mode of RowClone—the pipelined serial mode (PSM). Although not as fast or energy efficient as FPM, PSM addresses these two limitations of FPM.

5.2 Pipelined Serial Mode

The PSM efficiently copies data from a source row in one bank to a destination row in a *different* bank. PSM exploits the fact that a single internal bus that is shared across all the banks is used for both read and write operations. This enables the opportunity to copy an arbitrary quantity of data one cache line at a time from one bank to another in a pipelined manner.

To copy data from a source row in one bank to a destination row in a different bank, PSM first activates the corresponding rows in both banks. It then puts the source bank into *read mode*, the destination bank into *write mode*, and transfers data one cache line (corresponding to a column of data— 64 bytes) at a time. For this purpose, RowClone introduces a new DRAM command called TRANSFER. The TRANSFER command takes four parameters: (1) source bank index, (2) source column index, (3) destination bank index, and (4) destination column index. It copies the cache line corresponding to the source column index in the activated row of the source bank to the cache line corresponding to the destination column index in the activated row of the destination bank.

Unlike READ/WRITE, which interact with the memory channel connecting the processor and main memory, TRANSFER does not transfer data outside the chip. Fig. 15 pictorially compares the operation of the TRANSFER command with that of READ and WRITE. The thick (blue) lines indicate the data flow corresponding to the three commands. As shown in the figure, in contrast to the READ or WRITE commands, TRANSFER does not transfer data from or to the memory channel.

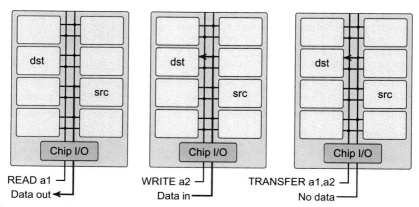

Fig. 15 RowClone: pipelined serial mode.

5.3 Mechanism for Bulk Data Copy

When the data from a source row (src) needs to be copied to a destination row (dst), there are three possible cases depending on the location of src and dst: (1) src and dst are within the same subarray, (2) src and dst are in different banks, and (3) src and dst are in different subarrays within the same bank. For case 1 and case 2, RowClone uses FPM and PSM, respectively, to complete the operation (as described in Sections 5.1 and 5.2).

For the case 3, when src and dst are in different subarrays within the same bank, one can imagine a mechanism that uses the global bitlines (shared across all subarrays within a bank—described in [48]) to copy data across the two rows in different subarrays. However, RowClone does not employ such a mechanism for two reasons. First, it is not possible in today's DRAM chips to activate multiple subarrays within the same bank simultaneously. Second, even if we enable simultaneous activation of multiple subarrays, as in [48], transferring data from one row buffer to another using the global bitlines requires the bank I/O circuitry to switch between read and write modes for each cache line transfer. This switching incurs significant latency overhead. To keep the design simple, for such an intra-bank copy operation, RowClone uses PSM to first copy the data from src to a temporary row (tmp) in a different bank. It then uses PSM again to copy the data from tmp to dst. The capacity lost due to reserving one row within each bank is negligible (0.0015% for a bank with 64 k rows).

Despite its location constraints, FPM can be used to accelerate *Copy-on-Write* (CoW), an important primitive in modern systems. CoW is used by most modern operating systems (OS) to postpone an expensive copy operation until it is actually needed. When data of one virtual page needs to be copied to another, instead of creating a copy, the OS points both virtual pages to the same physical page (source) and marks the page as read-only. In the future, when one of the sharers attempts to write to the page, the OS allocates a new physical page (destination) for the writer and copies the contents of the source page to the newly allocated page. Fortunately, prior to allocating the destination page, the OS already knows the location of the source physical page. Therefore, it can ensure that the destination is allocated in the same subarray as the source, thereby enabling the processor to use FPM to perform the copy.

5.4 Mechanism for Bulk Data Initialization

Bulk data initialization sets a large block of memory to a specific value. To perform this operation efficiently, RowClone first initializes a single DRAM row with the corresponding value. It then uses the appropriate copy mechanism (from Section 5.3) to copy the data to the other rows to be initialized.

Bulk zeroing (or BuZ), a special case of bulk initialization, is a frequently occurring operation in today's systems [65, 66]. To accelerate BuZ, one can reserve one row in each subarray that is always initialized to zero. By doing so, RowClone can use FPM to efficiently BuZ any row in DRAM by copying data from the reserved zero row of the corresponding subarray into the destination row. The capacity loss of reserving one row out of 512 rows in each subarray is very modest (0.2%).

While the reserved rows can potentially lead to gaps in the physical address space, we can use an appropriate memory interleaving technique that maps consecutive rows to different subarrays. Such a technique ensures that the reserved zero rows are contiguously located in the physical address space. Note that interleaving techniques commonly used in today's systems (e.g., row or cache line interleaving) have this property.

6. IN-DRAM BULK AND AND OR

In this section, we describe In-DRAM AND/OR (IDAO), which is a mechanism to perform bulk bitwise AND and OR operations completely inside DRAM. In addition to simple masking and initialization tasks, these operations are useful in important data structures like bitmap indices. For example, bitmap indices [67, 68] can be more efficient than commonly used B-trees for performing range queries and joins in databases [67, 69, 70]. In fact, bitmap indices are supported by many real-world database implementations (e.g., Redis [71] and Fastbit [69]). Improving the throughput of bitwise AND and OR operations can boost the performance of such bitmap indices.

6.1 Mechanism

As described in Section 4.2.2, when a DRAM cell is connected to a bitline precharged to $\frac{1}{2}V_{DD}$, the cell induces a deviation on the bitline, and the deviation is amplified by the sense amplifier. IDAO exploits the following fact about DRAM cell operation.

The final state of the bitline after amplification is determined solely by the deviation on the bitline after the charge sharing phase (after state ❸ in Fig. 6). If the deviation is positive (i.e., toward V_{DD}), the bitline is amplified to V_{DD}. Otherwise, if the deviation is negative (i.e., toward 0), the bitline is amplified to 0.

6.1.1 Triple-Row Activation

IDAO simultaneously connects three cells as opposed to a single cell to a sense amplifier. When three cells are connected to the bitline, the deviation of the bitline after charge sharing is determined by the *majority value* of the three cells. Specifically, if at least two cells are initially in the charged state, the effective voltage level of the three cells is at least $\frac{2}{3}V_{DD}$. This results in a positive deviation on the bitline. On the other hand, if at most one cell is initially in the charged state, the effective voltage level of the three cells is at most $\frac{1}{3}V_{DD}$. This results in a negative deviation on the bitline voltage. As a result, the final state of the bitline is determined by the logical majority value of the three cells.

Fig. 16 shows an example of activating three cells simultaneously. In the figure, we assume that two of the three cells are initially in the charged state and the third cell is in the empty state ❶. When the wordlines of all the three cells are raised simultaneously ❷, charge sharing results in a positive deviation on the bitline. Hence, after sense amplification, the sense amplifier drives the bitline to V_{DD} and fully charges all three cells ❸.

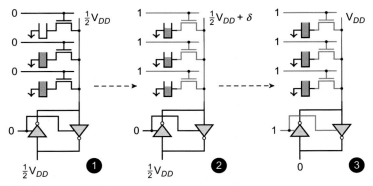

Fig. 16 Triple-row activation.

More generally, if the cell's capacitance is C_c, the bitline's is C_b, and if k of the three cells are initially in the charged state, based on the charge sharing principles [72], the deviation δ on the bitline voltage level is given by,

$$\delta = \frac{k \cdot C_c \cdot V_{DD} + C_b \cdot \frac{1}{2} V_{DD}}{3 C_c + C_b} - \frac{1}{2} V_{DD}$$
$$= \frac{(2k-3) C_c}{6 C_c + 2 C_b} V_{DD}$$

(1)

From the above equation, it is clear that δ is positive for $k = 2$, 3, and δ is negative for $k = 0$, 1. Therefore, after amplification, the final voltage level on the bitline is V_{DD} for $k = 2$, 3 and 0 for $k = 0$, 1.

If A, B, and C represent the logical values of the three cells, then the final state of the bitline is $AB + BC + CA$ (i.e., at least two of the values should be 1 for the final state to be 1). Importantly, using simple Boolean algebra, this expression can be rewritten as $C(A + B) + \overline{C}(AB)$. In other words, if the initial state of C is 1, then the final state of the bitline is a bitwise OR of A and B. Otherwise, if the initial state of C is 0, then the final state of the bitline is a bitwise AND of A and B. Therefore, by controlling the value of the cell C, we can execute a bitwise AND or bitwise OR operation of the remaining two cells using the sense amplifier. Due to the regular bulk operation of cells in DRAM, this approach naturally extends to an entire row of DRAM cells and sense amplifiers, enabling a multikilobyte-wide bitwise AND/OR operation.

6.1.2 Challenges
There are two challenges in this approach. First, Eq. (1) assumes that the cells involved in the triple-row activation are either fully charged or fully empty. However, DRAM cells leak charge over time. Therefore, the triple-row activation may not operate as expected. This problem may be exacerbated by process variation in DRAM cells. Second, as shown in Fig. 16 (state ❸), at the end of the triple-row activation, the data in all the three cells are overwritten with the final state of the bitline. In other words, this approach overwrites the source data with the final value. In the following sections, we describe a simple implementation of IDAO that addresses these challenges.

6.1.3 Implementation of IDAO
To ensure that the source data does not get modified, IDAO first *copies* the data from the two source rows to two reserved temporary rows ($T1$ and $T2$).

Depending on the operation to be performed (AND or OR), our mechanism initializes a third reserved temporary row $T3$ to (0 or 1). It then simultaneously activates the three rows $T1$, $T2$, and $T3$. It finally copies the result to the destination row. For example, to perform a bitwise AND of two rows A and B and store the result in row R, IDAO performs the following steps.

1. *Copy* data of row A to row $T1$
2. *Copy* data of row B to row $T2$
3. *Initialize* row $T3$ to 0
4. *Activate* rows $T1$, $T2$, and $T3$ simultaneously
5. *Copy* data of row $T1$ to row R

While the above mechanism is simple, the copy operations, if performed naively, will nullify the benefits of our mechanism. Fortunately, IDAO uses RowClone (described in Section 5), to perform row-to-row copy operations quickly and efficiently within DRAM. To recap, RowClone-FPM copies data within a subarray by issuing two back-to-back ACTIVATEs to the source row and the destination row, without an intervening PRECHARGE. RowClone-PSM efficiently copies data between two banks by using the shared internal bus to overlap the read to the source bank with the write to the destination bank.

With RowClone, all three copy operations (steps 1, 2, and 5) and the initialization operation (step 3) can be performed efficiently within DRAM. To use RowClone for the initialization operation, IDAO reserves two additional rows, $C0$ and $C1$. $C0$ is preinitialized to 0 and $C1$ is preinitialized to 1. Depending on the operation to be performed, our mechanism uses RowClone to copy either $C0$ or $C1$ to $T3$. Furthermore, to maximize the use of RowClone-FPM, IDAO reserves five rows in each subarray to serve as the temporary rows ($T1$, $T2$, and $T3$) and the control rows ($C0$ and $C1$).

In the best case, when all the three rows involved in the operation (A, B, and R) are in the same subarray, IDAO can use RowClone-FPM for all copy and initialization operations. However, if the three rows are in different banks/subarrays, some of the three copy operations have to use RowClone-PSM. In the worst case, when all three copy operations have to use RowClone-PSM, IDAO will consume higher latency than the baseline. However, when only one or two RowClone-PSM operations are required, IDAO will be faster and more energy efficient than existing systems. As our goal in this chapter is to demonstrate the power of our approach, in the rest of the chapter, we will focus our attention on the case when all rows involved in the bitwise operation are in the same subarray.

6.1.4 Reliability of Our Mechanism

While the above implementation trivially addresses the second challenge (modification of the source data), it also addresses the first challenge (DRAM cell leakage). This is because, in our approach, the source (and the control) data are copied to the rows $T1$, $T2$, and $T3$ *just* before the triple-row activation. Each copy operation takes much less than 1 μs, which is five *orders* of magnitude less than the typical refresh interval (64 ms). Consequently, the cells involved in the triple-row activation are very close to the fully refreshed state before the operation, thereby ensuring reliable operation of the triple-row activation. Having said that, an important aspect of the implementation is that a chip that fails the tests for triple-row activation (e.g., due to process variation) *can still be used as a regular DRAM chip*. As a result, this approach is likely to have little impact on the overall yield of DRAM chips, which is a major concern for manufacturers.

6.1.5 Latency Optimization

To complete an intra-subarray copy, RowClone-FPM uses two ACTIVATEs (back-to-back) followed by a PRECHARGE operation. Assuming typical DRAM timing parameters (tRAS = 35 ns and tRP = 15 ns), each copy operation consumes 85 ns. As IDAO is essentially four RowClone-FPM operations (as described in the previous section), the overall latency of a bitwise AND/OR operation is 4×85 ns = 340 ns.

In a RowClone-FPM operation, although the second ACTIVATE does not involve any sense amplification (the sense amplifiers are already activated), the RowClone paper [25] assumes the ACTIVATE consumes the full tRAS latency. However, by controlling the rows $D1$, $D2$, and $D3$ using a separate row decoder, it is possible to overlap the ACTIVATE to the destination fully with the ACTIVATE to the source row, by raising the wordline of the destination row toward the end of the sense amplification of the source row. This mechanism is similar to the inter-segment copy operation described in Tiered-Latency DRAM [49] (section 4.4). With this aggressive mechanism, the latency of a RowClone-FPM operation reduces to 50 ns (one ACTIVATE and one PRECHARGE). Therefore, the overall latency of a bitwise AND/OR operation is 200 ns. We will refer to this enhanced mechanism as *aggressive*, and the approach that uses the simple back-to-back ACTIVATE operations as *conservative*.

7. END-TO-END SYSTEM SUPPORT

Both RowClone and IDAO are substrates that exploit DRAM technology to perform bulk copy, initialization, and bitwise AND/OR operations efficiently inside DRAM. However, to exploit these substrates, we need support from the rest of the layers in the system stack, namely, the instruction set architecture, the microarchitecture, and the system software. In this section, we describe this support in detail.

7.1 ISA Support

To enable the software to communicate occurrences of the bulk operations to the hardware, the mechanisms introduce four new instructions to the Isa: memcopy, meminit, memand, and memor. Table 2 describes the semantics of these four new instructions. The mechanisms deliberately keep the semantics of the instructions simple in order to relieve the software from worrying about microarchitectural aspects of the DRAM substrate such as row size, alignment, etc. (discussed in Section 7.2.1). Note that such instructions are already present in some of the instructions sets in modern processors—e.g., rep movsd, rep stosb, ermsb in x86 [73], and mvcl in IBM S/390 [74].

There are three points to note regarding the execution semantics of these operations. First, the processor does not guarantee atomicity for any of these instructions, but note that existing systems also do not guarantee atomicity for such operations. Therefore, the software must take care of atomicity requirements using explicit synchronization. However, the microarchitectural implementation ensures that any data in the on-chip caches is kept consistent during the execution of these operations

Table 2 Semantics of the Memcopy, Meminit, Memand, and Memor Instructions

Instruction	Operands	Semantics
memcopy	*src, dst, size*	Copy *size* bytes from *src* to *dst*
meminit	*dst, size, val*	Set *size* bytes to *val* at *dst*
memand	*src1, src2, dst, size*	Perform bitwise AND of *size* bytes of *src1* with *size* bytes of *src2* and store the result in the *dst*
memor	*src1, src2, dst, size*	Perform bitwise OR of *size* bytes of *src1* with *size* bytes of *src2* and store the result in the *dst*

(Section 7.2.2). Second, the processor handles any page faults during the execution of these operations. Third, the processor can take interrupts during the execution of these operations.

7.2 Processor Microarchitecture Support

The microarchitectural implementation of the new instructions has two parts. The first part determines if a particular instance of the instructions can be fully/partially accelerated by RowClone/IDAO. The second part involves the changes required to the cache coherence protocol to ensure coherence of data in the on-chip caches. We discuss these parts in this section.

7.2.1 Source/Destination Alignment and Size

For the processor to accelerate a copy/initialization operation using RowClone, the operation must satisfy certain alignment and size constraints. Specifically, for an operation to be accelerated by FPM: (1) the source and destination regions should be within the same subarray, (2) the source and destination regions should be row-aligned, and (3) the operation should span an entire row. On the other hand, for an operation to be accelerated by PSM, the source and destination regions should be cache line-aligned and the operation must span a full cache line.

Upon encountering a memcopy/meminit instruction, the processor divides the region to be copied/initialized into three portions: (1) row-aligned row-sized portions that can be accelerated using FPM, (2) cache line-aligned cache line-sized portions that can be accelerated using PSM, and (3) the remaining portions that can be performed by the processor. For the first two regions, the processor sends appropriate requests to the memory controller, which completes the operations and sends an acknowledgment back to the processor. Since TRANSFER copies only a single cache line, a bulk copy using PSM can be interleaved with other commands to memory. The processor completes the operation for the third region similarly to how it is done in today's systems. Note that the CPU can offload all these operations to the memory controller. In such a design, the CPU need not be made aware of the DRAM organization (e.g., row size and alignment, subarray mapping).

For each instance of memand/memor instruction, the processor follows a similar procedure. However, only the row-aligned row-sized portions are accelerated using IDAO. The remaining portions are still performed by the CPU. For the row-aligned row-sized regions, some of the copy operations may require RowClone-PSM. For each row of the operation, the processor determines if the number of RowClone-PSM operations required is three. If so, the processor completes the execution in the CPU. Otherwise, the operation is completed using IDAO.

7.2.2 Managing On-Chip Cache Coherence

Both RowClone and IDAO allow the memory controller to directly read/modify data in memory without going through the on-chip caches. Therefore, to ensure cache coherence, the controller appropriately handles cache lines from the source and destination regions that may be present in the caches before issuing the in-DRAM operations to memory.

First, the memory controller writes back any dirty cache line from the source region as the main memory version of such a cache line is likely stale. Using the data in memory before flushing such cache lines will lead to stale data being copied to the destination region. Second, the controller invalidates any cache line (clean or dirty) from the destination region that is cached in the on-chip caches. This is because after performing the operation, the cached version of these blocks may contain stale data. The controller already has the ability to perform such flushes and invalidations to support direct memory access (DMA) [75]. After performing the necessary flushes and invalidations, the memory controller performs the in-DRAM operation. To ensure that cache lines of the destination region are not cached again by the processor in the meantime, the memory controller blocks all requests (including prefetches) to the destination region until the copy or initialization operation is complete. A recent work, LazyPIM [37], proposes an approach to perform the coherence operations lazily by comparing the signatures of data that were accessed in memory and the data that are cached on-chip. Our mechanisms can be combined with such works.

For RowClone, while performing the flushes and invalidates as mentioned above will ensure coherence, we propose a modified solution to handle dirty cache lines of the source region to reduce memory bandwidth consumption. When the memory controller identifies a dirty cache line belonging to the source region while performing a copy, it creates an in-cache copy of the source cache line with the tag corresponding to the destination cache line. This has two benefits. First, it avoids the additional memory flush required for the dirty source cache line. Second and more importantly, the controller does not have to wait for all the dirty source cache lines to be flushed before it can perform the copy. In the evaluation section, we will describe another optimization, called RowClone-Zero-Insert, which inserts clean zero cache lines into the cache to further optimize bulk zeroing. This optimization does not require further changes to the proposed modifications to the cache coherence protocol.

Although the two mechanisms require the controller to manage cache coherence, it does not affect memory consistency—i.e., the ordering of accesses by concurrent readers and/or writers to the source or destination regions. As mentioned before, such an operation is not guaranteed to be atomic even in current systems, and software needs to perform the operation within a critical section to ensure atomicity.

7.3 Software Support

The minimum support required from the system software is the use of the proposed instructions to indicate bulk data operations to the processor. Although one can have a working system with just this support, the maximum latency and energy benefits can be obtained if the hardware is able to accelerate most operations using FPM rather than PSM. Increasing the likelihood of the use of the FPM mode requires further support from the operating system (OS) on two aspects: (1) page mapping and (2) granularity of the operation.

7.3.1 Subarray-Aware Page Mapping

The use of FPM requires the source row and the destination row of a copy operation to be within the same subarray. Therefore, to maximize the use of FPM, the OS page mapping algorithm should be aware of subarrays so that it can allocate a destination page of a copy operation in the same subarray as the source page. More specifically, the OS should have knowledge of which pages map to the same subarray in DRAM. We propose that DRAM expose this information to software using the small electrically erasable programmable read-only memory (EEPROM) that already exists in today's DRAM modules. This EEPROM, called the serial presence detect (SPD) [76], stores information about the DRAM chips that is read by the memory controller at system bootup. Exposing the subarray mapping information will require only a few additional bytes to communicate the bits of the physical address that map to the subarray index. To increase DRAM yield, DRAM manufacturers design chips with spare rows that can be mapped to faulty rows [77]. The mechanisms can work with this technique by either requiring that each faulty row is remapped to a spare row within the same subarray, or exposing the location of all faulty rows to the memory controller so that it can use PSM to copy data across such rows.

Once the OS has the mapping information between physical pages and subarrays, it maintains multiple pools of free pages, one pool for each subarray. When the OS allocates the destination page for a copy operation (e.g., for a *Copy-on-Write* operation), it chooses the page from the same pool (subarray) as the source page. Note that this approach does not require contiguous pages to be placed within the same subarray. As mentioned before, commonly used memory interleaving techniques spread out contiguous pages across as many banks/subarrays as possible to improve parallelism. Therefore, both the source and destination of a bulk copy operation can be spread out across many subarrays.

7.3.2 Granularity of the Operations

The second aspect that affects the use of FPM and IDAO is the granularity at which data are copied or initialized. These mechanisms have a minimum granularity at which they operate. There are two factors that affect this minimum granularity: (1) the size of each DRAM row and (2) the memory interleaving employed by the controller.

First, in each chip, these mechanisms operate on an entire row of data. Second, to extract maximum bandwidth, some memory interleaving techniques map consecutive cache lines to different memory channels in the system. Therefore, to operate on a contiguous region of data with such interleaving strategies, the mechanisms must perform the operation in each channel. The minimum amount of data in such a scenario is the product of the row size and the number of channels.

To maximize the likelihood of using FPM and IDAO, the system or application software must ensure that the region of data involved in the operation is at least as large as this minimum granularity. For this purpose, we propose to expose this minimum granularity to the software through a special register, which we call the *Minimum DRAM Granularity Register* (MDGR). On system bootup, the memory controller initializes the MCGR based on the row size and the memory interleaving strategy, which can later be used by the OS for effectively exploiting RowClone/IDAO. Note that some previously proposed techniques such as subwordline activation [62] or mini-rank [78, 79] can be combined with our mechanisms to reduce the minimum granularity.

8. EVALUATION

To highlight the benefits of performing various operations completely inside DRAM, we first compare the raw latency and energy required to perform these operations using different mechanisms (Section 8.1). We then present the quantitative evaluation of applications for RowClone and IDAO in Sections 8.2 and 8.3, respectively.

8.1 Latency and Energy Analysis

We estimate latency using DDR3-1600 timing parameters. We estimate energy using the Rambus power model [61]. Our energy calculations only include the energy consumed by the DRAM module and the DRAM channel. Table 3 shows the latency and energy consumption due to the different mechanisms for bulk copy, zero, and bitwise AND/OR operations. The table also shows the potential reduction in latency and energy by performing these operations completely inside DRAM.

First, for bulk copy operations, RowClone-FPM reduces latency by 12× and energy consumption by 74.4× compared to existing interfaces. While PSM does not provide as much reduction as FPM, PSM still reduces latency by 2× and energy consumption by 3.2× compared to the baseline. Second,

Table 3 DRAM Latency and Memory Energy Reductions

	Mechanism	Absolute		Reduction	
		Latency (ns)	Memory Energy (μJ)	Latency	Memory Energy
Copy	Baseline	1020	3.6	1.00×	1.0×
	FPM	85	0.04	12.0×	74.4×
	Inter–Bank—PSM	510	1.1	2.0×	3.2×
	Intra–Bank—PSM	1020	2.5	1.0×	1.5×
Zero	Baseline	510	2.0	1.00×	1.0×
	FPM	85	0.05	6.0×	41.5×
AND/OR	Baseline	1530	5.0	1.00×	1.0×
	IDAO–Conservative	320	0.16	4.78×	31.6×
	IDAO–Aggressive	200	0.10	7.65×	50.5×

for bulk zeroing operations, RowClone can always use the FPM mode as it reserves a single zero row in each subarray of DRAM. As a result, it can reduce the latency of bulk zeroing by 6× and energy consumption by 41.5× compared to existing interfaces. Finally, for bitwise AND/OR operations, even with conservative timing parameters, IDAO can reduce latency by 4.78× and energy consumption by 31.6×. With more aggressive timing parameters, IDAO reduces latency by 7.65× and energy by 50.5×.

The improvement in sustained throughput due to RowClone and IDAO for the respective operations is similar to the improvements in latency. The main takeaway from these results is that, for systems that use DRAM to store majority of their data (which includes most of today's systems), these in-DRAM mechanisms are probably the best performing and the most energy efficient way of performing the respective operations. We will now provide quantitative evaluation of these mechanisms on some real applications.

8.2 Applications for RowClone

Our evaluations use an in-house cycle-level multicore simulator similar to memsim [80–82] along with a cycle-accurate command-level DDR3 DRAM simulator, similar to Ramulator [53, 83]. The multicore simulator models out-of-order cores, each with a private last-level cache. We evaluate the benefits of RowClone using (1) a case study of the fork system call, an important operation used by modern operating systems, (2) six copy and initialization-intensive benchmarks: *bootup*, *compile*, *forkbench*, *memcached* [84], *mysql* [85], and *shell* (Section 8.2.2 describes these benchmarks), and (3) a wide variety of multicore workloads comprising the copy/initialization-intensive applications running alongside memory-intensive applications from the SPEC CPU2006 benchmark suite [86]. Note that benchmarks such as SPEC CPU2006, which predominantly stress the CPU, typically use a small number of page copy and initialization operations and therefore would serve as poor individual evaluation benchmarks for RowClone.

We collected instruction traces for our workloads using Bochs [87], a full system x86-64 emulator, running a GNU/Linux system. We modify the kernel's implementation of page copy/initialization to use the memcopy and meminit instructions and mark these instructions in our traces. For the fork benchmark, we used the Wind River Simics full system simulator [88] to collect the traces. We collect 1-billion instruction traces of the representative portions of these workloads. We use the instruction throughput (IPC) metric to measure single-core performance. We evaluate multicore runs using the weighted speedup metric [89, 90]. This metric is used by many prior works (e.g., [91–98]) to measure system throughput for multiprogrammed workloads. In addition to weighted speedup, we use five other performance/fairness/bandwidth/energy metrics, as shown in Table 7.

8.2.1 The fork System Call

fork is one of the most expensive yet frequently used system calls in modern systems [99]. Since fork triggers a large number of CoW operations (as a result of updates to shared pages from the parent or child process), RowClone can significantly improve the performance of fork.

The performance of fork depends on two parameters: (1) the size of the address space used by the parent—which determines how much data may potentially have to be copied, and (2) the number of pages updated after the fork operation by either the parent or the child—which determines how much data are actually copied. To exercise these two parameters, we create a microbenchmark, forkbench, which first creates an array of size S and initializes the array with random values. It then forks itself. The child process updates N random pages (by updating a cache line within each page) and exits; the parent process waits for the child process to complete before exiting itself.

As such, we expect the number of copy operations to depend on N—the number of pages copied. Therefore, one may expect RowClone's performance benefits to be proportional to N. However, an application's performance typically depends on the *overall memory access rate* [100, 101], and RowClone can only improve performance by reducing the *memory access rate due to copy operations*. As a result, we expect the performance improvement due to RowClone to primarily depend on the *fraction* of memory traffic (total bytes transferred over the memory channel) generated by copy operations. We refer to this fraction as FMTC—fraction of memory traffic due to copies.

Fig. 17 plots FMTC of forkbench for different values of S (64 MB and 128 MB) and N (2–16 k) in the baseline system. As the figure shows, for both values of S, FMTC increases with increasing N. This is expected as a higher N (more pages updated by the child) leads to more CoW operations. However, because of the presence of other read/write operations (e.g., during the initialization phase of the parent), for a given value of N,

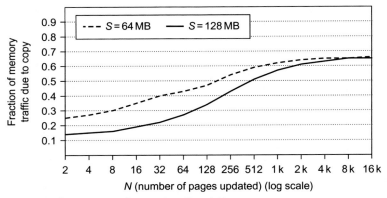

Fig. 17 FMTC of forkbench for varying S and N.

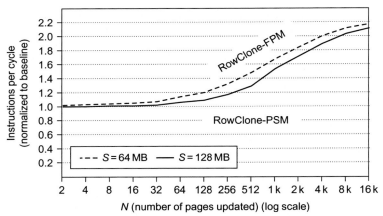

Fig. 18 Performance improvement due to RowClone for forkbench with different values of S and N.

FMTC is larger for $S = 64$ MB compared to $S = 128$ MB. Depending on the value of S and N, anywhere between 14% and 66% of the memory traffic arises from copy operations. This shows that accelerating copy operations using RowClone has the potential to significantly improve the performance of the fork operation.

Fig. 18 plots the performance (IPC) of FPM and PSM for forkbench, normalized to that of the baseline system. We draw two conclusions from the figure. First, FPM improves the performance of forkbench for both values of S and most values of N. The peak performance improvement is $2.2\times$ for $N = 16$ k (30% on average across all data points). As expected, the improvement of FPM increases as the number of pages updated increases. The trend in performance improvement of FPM is similar to that of FMTC (Fig. 17), confirming our hypothesis that FPM's performance improvement primarily depends on FMTC. Second, PSM does not provide considerable performance improvement over the baseline. This is because the large on-chip cache in the baseline system buffers the writebacks generated by the copy operations. These writebacks are flushed to memory at a later point without further delaying the copy operation. As a result, PSM, which just overlaps the read and write operations involved in the copy, does not improve latency significantly in the presence of a large on-chip cache. On the other hand, FPM, by copying all cache lines from the source row to destination in parallel, significantly reduces the latency compared to the baseline (which still needs to read the source blocks from main memory), resulting in high-performance improvement.

Fig. 19 shows the reduction in DRAM energy consumption (considering both the DRAM and the memory channel) of FPM and PSM modes of RowClone compared to that of the baseline for forkbench with $S = 64$ MB. Similar to performance, the overall DRAM energy consumption also depends on the total memory access rate. As a result, RowClone's potential to reduce DRAM energy depends on the fraction of memory traffic generated by copy operations. In fact, our results also show that the DRAM energy reduction due to FPM and PSM correlate well with FMTC (Fig. 17). By efficiently performing the copy operations, FPM reduces DRAM energy consumption by up to 80% (average 50%, across

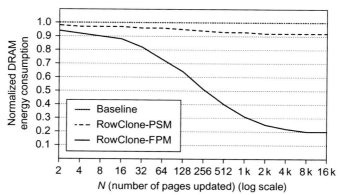

Fig. 19 Comparison of DRAM energy consumption of different mechanisms for forkbench ($S = 64$ MB).

all data points). Similar to FPM, the energy reduction of PSM also increases with increasing N with a maximum reduction of 9% for $N = 16$ k.

In a system that is agnostic to RowClone, we expect the performance improvement and energy reduction of RowClone to be in between that of FPM and PSM. By making the system software aware of RowClone (Section 7.3), i.e., designing the system software to be aware of the topology (subarray and bank organization) of DRAM, as also advocated by various recent works [102–104], we can approximate the maximum performance and energy benefits by increasing the likelihood of the use of FPM.

8.2.2 Copy/Initialization-Intensive Applications

In this section, we analyze the benefits of RowClone on six copy/initialization-intensive applications, including one instance of the forkbench described in the previous section. Table 4 describes these applications.

Fig. 20 plots the FMTC, initialization, and regular read/write operations for the six applications. For these applications, between 10% and 80% of the memory traffic is generated by copy and initialization operations.

Fig. 21 compares the IPC of the baseline with that of RowClone and a variant of RowClone, RowClone-ZI (described shortly). The RowClone-based initialization mechanism slightly degrades performance for the applications which have a negligible number of copy operations (*mcached*, *compile*, and *mysql*).

Our further analysis indicated that, for these applications, although the operating system zeroes out any newly allocated page, the application

Table 4 Copy/Initialization-Intensive Benchmarks

Name	Description
bootup	A phase booting up the Debian operating system.
compile	The compilation phase from the GNU C compiler (while running *cc1*).
forkbench	An instance of the forkbench described in Section 8.2.1 with $S = 64$ MB and $N = 1$ k.
mcached	Memcached [84], a memory object caching system, a phase inserting many key-value pairs into the memcache.
mysql	MySQL [85], an on-disk database system, a phase loading the sample *employeedb*
shell	A Unix shell script running 'find' on a directory tree with 'ls' on each subdirectory (involves file system accesses and spawning new processes).

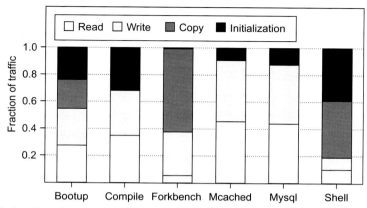

Fig. 20 Fraction of memory traffic due to read, write, copy, and initialization.

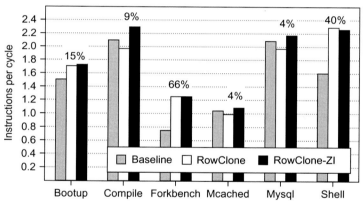

Fig. 21 Performance improvement of RowClone and RowClone-ZI. Value on top indicates percentage improvement of RowClone-ZI over baseline.

typically accesses almost all cache lines of a page immediately after the page is zeroed out. There are two phases: (1) the phase when the OS zeroes out the page, and (2) the phase when the application accesses the cache lines of the page. While the baseline incurs cache misses during phase 1, RowClone, as a result of performing the zeroing operation completely in memory, incurs cache misses in phase 2. However, the baseline zeroing operation is heavily optimized for memory-level parallelism (MLP) [105–109]. Memory-level parallelism indicates the number of concurrent outstanding misses to main memory. Higher MLP results in higher overlap in the latency of the requests. Consequently, higher MLP results in lower overall latency. In contrast, the cache misses in phase 2 have low MLP. As a result, incurring the same misses in phase 2 (as with RowClone) causes higher overall stall time for the application (because the latencies for the misses are serialized) than incurring them in phase 1 (as in the baseline), resulting in RowClone's performance degradation compared to the baseline.

To address this problem, RowClone uses a variant called RowClone-Zero-Insert (RowClone-ZI). RowClone-ZI not only zeroes out a page in DRAM but it also inserts a zero cache line into the processor cache corresponding to each cache line in the page that is zeroed out. By doing so, RowClone-ZI avoids the cache misses during both phase 1 (zeroing operation) and phase 2 (when the application accesses the cache lines of the zeroed page). As a result, it improves performance for all benchmarks, notably `forkbench` (by 66%) and *shell* (by 40%), compared to the baseline.

Table 5 shows the percentage reduction in DRAM energy and memory bandwidth consumption with RowClone and RowClone-ZI compared to the baseline. While RowClone significantly reduces both energy and memory bandwidth consumption for *bootup*, *forkbench*, and *shell*, it has negligible impact on both metrics for the remaining three benchmarks. The lack of energy and bandwidth benefits in these three applications is due to serial execution caused by the cache misses incurred when the processor accesses the zeroed out pages (i.e., *phase 2*, as described above), which also leads to performance degradation in these workloads (as also described above). RowClone-ZI, which eliminates the cache misses in *phase 2*, significantly reduces energy consumption (between 15% and 69%) and memory bandwidth consumption (between

Table 5 DRAM Energy and Bandwidth Reduction due to RowClone and RowClone-ZI (Indicated as +ZI)

Application	Energy Reduction		Bandwidth Reduction	
	RowClone	+ZI	RowClone	+ZI
bootup	39%	40%	49%	52%
compile	−2%	32%	2%	47%
forkbench	69%	69%	60%	60%
mcached	0%	15%	0%	16%
mysql	−1%	17%	0%	21%
shell	68%	67%	81%	81%

16% and 81%) for all benchmarks compared to the baseline. We conclude that RowClone-ZI can effectively improve performance, memory energy, and memory bandwidth efficiency in page copy and initialization-intensive single-core workloads.

8.2.3 Multicore Evaluations

As RowClone performs bulk data operations completely within DRAM, it significantly reduces the memory bandwidth consumed by these operations. As a result, RowClone can benefit other applications running concurrently on the same system. We evaluate this benefit of RowClone by running our copy/initialization-intensive applications alongside memory-intensive applications from the SPEC CPU2006 benchmark suite [86] (i.e., those applications with last-level cache MPKI greater than 1). Table 6 lists the set of applications used for our multiprogrammed workloads.

We generate multiprogrammed workloads for 2-core, 4-core, and 8-core systems. In each workload, half of the cores run copy/initialization-intensive benchmarks and the remaining cores run memory-intensive SPEC benchmarks. Benchmarks from each category are chosen at random.

Fig. 22 plots the performance improvement due to RowClone and RowClone-ZI for the 50 4-core workloads we evaluated (sorted based on the performance improvement due to RowClone-ZI). Two conclusions are in order. First, although RowClone degrades performance of certain 4-core workloads (with *compile*, *mcached*, or *mysql* benchmarks), it significantly improves performance for all other workloads (by 10% across all workloads). Second, like in our single-core evaluations (Section 8.2.2), RowClone-ZI eliminates the performance degradation due to RowClone and consistently outperforms both the baseline and RowClone for all workloads (20% on average).

Table 6 List of Benchmarks Used for Multicore Evaluation

Copy/initialization-intensive benchmarks
bootup, compile, forkbench, mcached, mysql, shell
Memory-intensive benchmarks from SPEC CPU2006
bzip2, gcc, mcf, milc, zeusmp, gromacs, cactusADM, leslie3d, namd, gobmk, dealII, soplex, hmmer, sjeng, GemsFDTD, libquantum, h264ref, lbm, omnetpp, astar, wrf, sphinx3, xalancbmk

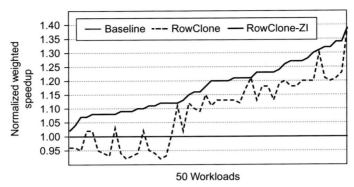

Fig. 22 System performance improvement of RowClone for 4-core workloads.

Table 7 Multicore Performance, Fairness, Bandwidth, and Energy

Number of Cores	2	4	8
Number of workloads	138	50	40
Weighted speedup [89, 90] improvement	15%	20%	27%
Instruction throughput improvement	14%	15%	25%
Harmonic speedup [125] improvement	13%	16%	29%
Maximum slowdown [110–112] reduction	6%	12%	23%
Memory bandwidth/instruction [126] reduction	29%	27%	28%
Memory energy/instruction reduction	19%	17%	17%

Table 7 shows the number of workloads and six metrics that evaluate the performance, fairness, memory bandwidth, and energy efficiency improvement due to RowClone compared to the baseline for systems with 2, 4, and 8 cores. We evaluate fairness using the maximum slowdown metric, which has been used by many prior works [80, 93, 94, 100, 101, 110–124] as an indicator of unfairness in the system. Maximum slowdown is defined as the maximum of the slowdowns of all applications that are in the multicore workload. For all three systems, RowClone significantly outperforms the baseline on all metrics.

To provide more insight into the benefits of RowClone on multicore systems, we classify our copy/initialization–intensive benchmarks into two categories: (1) Moderately copy/initialization-intensive (*compile*, *mcached*, and *mysql*) and highly copy/initialization-intensive (*bootup*, *forkbench*, and *shell*). Fig. 23 shows the average improvement in weighted speedup for

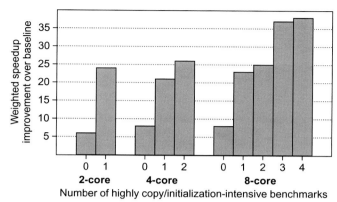

Fig. 23 Effect of increasing copy/initialization intensity.

the different multicore workloads, categorized based on the number of highly copy/initialization-intensive benchmarks. As the trends indicate, the performance improvement increases with increasing number of such benchmarks for all three multicore systems, indicating the effectiveness of RowClone in accelerating bulk copy/initialization operations.

We conclude that RowClone is an effective mechanism to improve system performance, energy efficiency and bandwidth efficiency of future, memory-bandwidth-constrained multicore systems.

8.2.4 Memory-Controller-Based DMA

One alternative way to perform a bulk data operation is to use the memory controller to complete the operation using the regular DRAM interface (similar to some prior approaches [65, 127]). We refer to this approach as the memory controller-based DMA (MC-DMA). MC-DMA can potentially avoid the cache pollution caused by inserting blocks (involved in the copy/initialization) unnecessarily into the caches. However, it still requires data to be transferred over the memory bus. Hence, it suffers from the large latency, bandwidth, and energy consumption associated with the data transfer. Because the applications used in our evaluations do not suffer from cache pollution, we expect MC-DMA to perform comparably or worse than the baseline. In fact, our evaluations show that MC-DMA degrades performance compared to our baseline by 2% on average for the six copy/initialization-intensive applications (16% compared to RowClone). In addition, MC-DMA does not conserve any DRAM energy, unlike RowClone.

8.2.5 Other Applications

8.2.5.1 Secure Deallocation

Most operating systems (e.g., Linux [128], Windows [129], and Mac OS X [130]) zero out pages newly allocated to a process. This is done to prevent malicious processes from gaining access to the data that previously belonged to other processes or the kernel itself. Not doing so can potentially lead to security vulnerabilities, as shown by prior works [131–134].

8.2.5.2 Process Checkpointing

Checkpointing is an operation during which a consistent version of a process state is backed-up, so that the process can be restored from that state in the future. This checkpoint-restore primitive is useful in many cases including high-performance computing servers [135], software debugging with reduced overhead [136], hardware-level fault and bug tolerance mechanisms [137–139], mechanisms to provide consistent updates of persistent memory state [140], and speculative OS optimizations to improve performance [141, 142]. However, to ensure that the checkpoint is consistent (i.e., the original process does not update data while the checkpointing is in progress), the pages of the process are marked with copy-on-write. As a result, check-pointing often results in a large number of CoW operations.

8.2.5.3 Virtual Machine Cloning/Deduplication

Virtual machine (VM) cloning [143] is a technique to significantly reduce the startup cost of VMs in a cloud computing server. Similarly, dedupli-cation is a technique employed by modern hypervisors [144] to reduce the overall memory capacity requirements of VMs. With this technique, different VMs share physical pages that contain the same data. Similar to forking, both these operations likely result in a large number of CoW oper-ations for pages shared across VMs.

8.2.5.4 Page Migration

Bank conflicts, i.e., concurrent requests to different rows within the same bank, typically result in reduced row buffer hit rate and hence degrade both system performance and energy efficiency. Prior work [145] proposed tech-niques to mitigate bank conflicts using page migration. The PSM mode of RowClone can be used in conjunction with such techniques to (1) signif-icantly reduce the migration latency and (2) make the migrations more energy efficient.

8.2.5.5 CPU-GPU Communication

In many current and future processors, the GPU is or is expected to be integrated on the same chip with the CPU. Even in such systems where the CPU and GPU share the same off-chip memory, the off-chip memory is partitioned between the two devices. As a consequence, whenever a CPU program wants to offload some computation to the GPU, it has to copy all the necessary data from the CPU address space to the GPU address space [146]. When the GPU computation is finished, all the data needs to be copied back to the CPU address space. This copying involves a significant overhead. In fact, a recent work, decoupled DMA [147], motivates this problem and proposes a solution to mitigate it. By spreading out the GPU address space over all subarrays and mapping the application data appropriately, RowClone can significantly speed up these copy operations. Note that communication between different processors and accelerators in a heterogeneous system-on-a-chip (SoC) is done similarly to the CPU-GPU communication and can also be accelerated by RowClone.

8.3 Applications for IDAO

We analyze our mechanism's performance on a real-world bitmap index library, FastBit [69], widely used in physics simulations and network analysis. Fastbit can enable faster and more efficient searching/retrieval compared to B-trees.

To construct an index, FastBit uses multiple bitmap bins, each corresponding to either a distinct value or a range of values. FastBit relies on fast bitwise AND/OR operations on these bitmaps to accelerate *joins* and *range queries*. For example, to execute a range query, FastBit performs a bitwise OR of all bitmaps that correspond to the specified range.

We initialized FastBit on our baseline system using the sample *STAR* [148] data set. We then ran a set of indexing-intensive range queries that touch various numbers of bitmap bins. For each query, we measure the fraction of query execution time spent on bitwise OR operations. Table 8 shows the corresponding results. For each query, the table shows the number of bitmap bins involved in the query and the percentage of time spent in

Table 8 Fraction of Time Spent in OR Operations

Number of Bins	3	9	20	45	98	118	128
Fraction of time spent in OR	29%	29%	31%	32%	34%	34%	34%

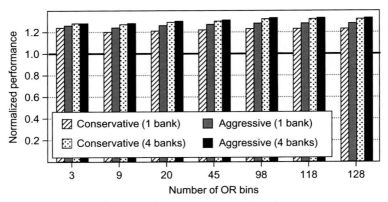

Fig. 24 Range query performance improvement over baseline.

bitwise OR operations. On average, 31% of the query execution is spent on bitwise OR operations (with small variance across queries).

To estimate the performance of our mechanism, we measure the number of bitwise OR operations required to complete the query. We then compute the amount of time taken by our mechanism to complete these operations and then use that to estimate the performance of the overall query execution. To perform a bitwise OR of more than two rows, our mechanism is invoked two rows at a time. Fig. 24 shows the potential performance improvement using our two mechanisms (conservative and aggressive), each with either 1 bank or 4 banks.

As our results indicate, our aggressive mechanism with 4 banks improves the performance of range queries by 30% (on average) compared to the baseline, eliminating almost all the overhead of bitwise operations. As expected, the aggressive mechanism performs better than the conservative mechanism. Similarly, using more banks provides better performance. Even if we assume a 2× higher latency for the triple-row activation, our conservative mechanism with 1 bank improves performance by 18% (not shown in the figure).

8.4 Recent Works Building on RowClone and IDAO

Some recent works [38, 149] have built on top of RowClone and IDAO and have proposed mechanisms to perform bulk copy and bitwise operations inside memory. As we described in Section 5.3, to copy data across two sub-arrays in the same bank, RowClone uses two PSM operations. While this approach reduces energy consumption compared to existing systems, it still does not reduce latency. Low-cost interlinked subarrays or LISA [149] addresses this problem by connecting adjacent subarrays of a bank. LISA exploits the open bitline architecture and connects the open end of each bitline to the adjacent sense amplifier using a transistor. LISA uses these connections to transfer data more efficiently and quickly across subarrays in the same bank. Pinatubo [38] takes an approach similar to IDAO and uses PCM technology to perform bitwise operations inside a memory chip built using PCM. Pinatubo enables the PCM sense amplifier to detect fine-grained differences in cell resistance. With the enhanced sense amplifier, Pinatubo can perform bitwise AND/OR operations by simultaneously sensing multiple PCM cells connected to the same sense amplifier.

9. CONCLUSION

In this chapter, we focused our attention on the problem of data movement, especially for operations that access a large amount of data. We first discussed the general notion of PiM as a potential solution to reducing data movement so as to achieve better performance and efficiency. PiM adds new logic structures, sometimes as large as simple processors, near memory, to perform computation. We then introduced the idea of PuM, which exploits some of the peripheral structures *already existing* inside memory devices (with minimal changes), to perform other tasks on top of storing data. PuM is a cost-effective approach as it does not add significant logic structures near or inside memory.

We developed two new ideas that take the PuM approach and build on top of DRAM technology. The first idea is RowClone, which exploits the underlying operation of DRAM to perform bulk copy and initialization operations completely inside DRAM. RowClone exploits the fact DRAM cells internally share several buses that can act as a fast path for copying data across them. The second idea is In-DRAM AND/OR (IDAO), which exploits the analog operation of DRAM to perform bulk bitwise AND/OR operations completely inside DRAM. IDAO exploits the fact that many DRAM cells share the same sense amplifier and uses simultaneous activation of three rows of DRAM cells to perform bitwise AND/OR operations efficiently.

Our evaluations show that both mechanisms (RowClone and IDAO) improve the performance and energy efficiency of the respective operations by more than an order of magnitude. In fact, for systems that store data in DRAM, these mechanisms are probably as efficient as any mechanism could be (since they minimize the amount of data movement). We described many real-world applications that can exploit RowClone and IDAO and have demonstrated significant performance and energy efficiency improvements using these mechanisms. Due to its low cost of implementation and large performance and energy benefits, we believe PuM is a very promising and viable approach to minimize the memory bottleneck in data-intensive applications. We hope and expect future research will build upon this approach to demonstrate other techniques that can perform more operations in memory.

REFERENCES

[1] W. Dally, GPU Computing to Exascale and Beyond, Invited talk at International Conference on High Performance Computing, Networking, Storage, and Analysis, 2010, http://www.nvidia.com/content/PDF/sc_2010/theater/Dally_SC10.pdf.

[2] J. Ahn, S. Yoo, O. Mutlu, K. Choi, PIM-enabled instructions: a low-overhead, locality-aware processing-in-memory architecture, in: Proceedings of the 42nd Annual International Symposium on Computer Architecture, ISCA '15, Portland, Oregon, ACM, New York, NY, ISBN 978-1-4503-3402-0, 2015, pp. 336–348, http://dx.doi.org/10.1145/2749469.2750385.

[3] J. Ahn, S. Hong, S. Yoo, O. Mutlu, K. Choi, A scalable processing-in-memory accelerator for parallel graph processing, in: Proceedings of the 42nd Annual International Symposium on Computer Architecture, ISCA '15, Portland, Oregon, ACM, New York, NY, ISBN 978-1-4503-3402-0, 2015, pp. 105–117, http://dx.doi.org/10.1145/2749469.2750386.

[4] D. Zhang, N. Jayasena, A. Lyashevsky, J.L. Greathouse, L. Xu, M. Ignatowski, TOP-PIM: throughput-oriented programmable processing in memory, in: Proceedings of the 23rd International Symposium on High-performance Parallel and Distributed Computing, HPDC '14, Vancouver, BC, Canada, ACM, New York, NY, ISBN 978-1-4503-2749-7, 2014, pp. 85–98, http://dx.doi.org/10.1145/2600212.2600213.

[5] A. Farmahini-Farahani, J.H. Ahn, K. Morrow, N.S. Kim, NDA: near-DRAM acceleration architecture leveraging commodity DRAM devices and standard memory modules, in: IEEE 21st International Symposium on High Performance Computer Architecture (HPCA), 2015, 2015, pp. 283–295, http://dx.doi.org/10.1109/HPCA.2015.7056040.

[6] Q. Guo, N. Alachiotis, B. Akin, F. Sadi, G. Xu, T.M. Low, L. Pileggi, J.C. Hoe, F. Frachetti, 3D-stacked memory-side acceleration: accelerator and system design, in: WoNDP, 2013.

[7] Q. Zhu, T. Graf, H.E. Sumbul, L. Pileggi, F. Franchetti, Accelerating sparse matrix-matrix multiplication with 3D-stacked logic-in-memory hardware, in: High Performance Extreme Computing Conference (HPEC), 2013 IEEE, 2013, pp. 1–6, http://dx.doi.org/10.1109/HPEC.2013.6670336.

[8] Z. Sura, A. Jacob, T. Chen, B. Rosenburg, O. Sallenave, C. Bertolli, S. Antao, J. Brunheroto, Y. Park, K. O'Brien, R. Nair, Data access optimization in a processing-in-memory system, in: Proceedings of the 12th ACM International Conference on Computing Frontiers, CF '15, Ischia, Italy, ACM, New York, NY, ISBN 978-1-4503-3358-0, 2015, pp. 6:1–6:8, http://dx.doi.org/10.1145/2742854.2742863.

[9] K. Hsieh, E. Ebrahimi, G. Kim, N. Chatterjee, M. O'Conner, N. Vijaykumar, O. Mutlu, S.W. Keckler, Transparent offloading and mapping (TOM): enabling programmer-transparent near-data processing in GPU systems, in: Proceedings of the 38th Annual International Symposium on Computer Architecture (ISCA), 2016.

[10] M. Gao, C. Kozyrakis, HRL: efficient and flexible reconfigurable logic for near-data processing, in: HPCA, 2016.

[11] A. Morad, L. Yavits, R. Ginosar, GP-SIMD processing-in-memory, ACM Trans. Archit. Code Optim. 11 (4) (2015) 53:1–53:26, ISSN 1544-3566, http://dx.doi.org/10.1145/2686875.

[12] S.M. Hassan, S. Yalamanchili, S. Mukhopadhyay, Near data processing: impact and optimization of 3D memory system architecture on the uncore, in: Proceedings of the 2015 International Symposium on Memory Systems, MEMSYS '15, Washington DC, USA, ACM, New York, NY, ISBN 978-1-4503-3604-8, 2015, pp. 11–21, http://dx.doi.org/10.1145/2818950.2818952.

[13] M. Gao, G. Ayers, C. Kozyrakis, Practical near-data processing for in-memory analytics frameworks, in: Proceedings of the 2015 International Conference on Parallel Architecture and Compilation (PACT), PACT '15, IEEE Computer Society, Washington, DC, ISBN 978-1-4673-9524-3, 2015, pp. 113–124, http://dx.doi.org/10.1109/PACT.2015.22.

[14] O. Babarinsa, S. Idreos, JAFAR: near-data processing for databases, in: Proceedings of the 2015 ACM SIGMOD International Conference on Management of Data, SIGMOD '15, Melbourne, Victoria, Australia, ACM, New York, NY, ISBN 978-1-4503-2758-9, 2015, pp. 2069–2070, http://dx.doi.org/10.1145/2723372.2764942.

[15] B. Akin, F. Franchetti, J.C. Hoe, Data reorganization in memory using 3D-stacked DRAM, in: Proceedings of the 42nd Annual International Symposium on Computer Architecture, ISCA '15, Portland, Oregon, ACM, New York, NY, ISBN 978-1-4503-3402-0, 2015, pp. 131–143, http://dx.doi.org/10.1145/2749469.2750397.

[16] D. Lee, S. Ghose, G. Pekhimenko, S. Khan, O. Mutlu, Simultaneous multi-layer access: improving 3D-stacked memory bandwidth at low cost, ACM Trans. Archit. Code Optim. 12 (4) (2016) 63:1–63:29, ISSN 1544-3566, http://dx.doi.org/10.1145/2832911.

[17] H.S. Stone, A logic-in-memory computer, IEEE Trans. Comput. 19 (1) (1970) 73–78, ISSN 0018-9340, http://dx.doi.org/10.1109/TC.1970.5008902.

[18] D.E. Shaw, S. Stolfo, H. Ibrahim, B.K. Hillyer, J. Andrews, G. Wiederhold, The NON-VON Database Machine: An Overview, 1981, Columbia University Academic Commons, http://hdl.handle.net/10022/AC:P:11530.

[19] D. Patterson, T. Anderson, N. Cardwell, R. Fromm, K. Keeton, C. Kozyrakis, R. Thomas, K. Yelick, A case for intelligent RAM, IEEE Micro 17 (2) (1997) 34–44, ISSN 0272-1732, http://dx.doi.org/10.1109/40.592312.

[20] P.M. Kogge, EXECUBE: a new architecture for scalable MPPs, in: Proceedings of the 1994 International Conference on Parallel Processing (ICPP), vol. 1, IEEE Computer Society, Washington, DC, ISBN 0-8493-2493-9, 1994, pp. 77–84, http://dx.doi.org/10.1109/ICPP.1994.108.

[21] M. Oskin, F.T. Chong, T. Sherwood, Active pages: a computation model for intelligent memory, in: Proceedings of the 25th Annual International Symposium on Computer Architecture, ISCA '98, Barcelona, Spain, IEEE Computer Society, Washington, DC, ISBN 0-8186-8491-7, 1998, pp. 192–203, http://dx.doi.org/10.1145/279358.279387.

[22] M. Gokhale, B. Holmes, K. Iobst, Processing in memory: the terasys massively parallel PIM array, Computer 28 (4) (1995) 23–31, ISSN 0018-9162, http://dx.doi.org/10.1109/2.375174.

[23] D. Elliott, M. Stumm, W.M. Snelgrove, C. Cojocaru, R. McKenzie, Computational RAM: implementing processors in memory, IEEE Des. Test Comput. 16 (1) (1999) 32–41, ISSN 0740-7475, http://dx.doi.org/10.1109/54.748803.

[24] V. Seshadri, K. Hsieh, A. Boroumand, D. Lee, M.A. Kozuch, O. Mutlu, P.B. Gibbons, T.C. Mowry, Fast bulk bitwise AND and OR in DRAM, IEEE

Comput. Archit. Lett. 14 (2) (2015) 127–131, ISSN 1556-6056, http://dx.doi.org/10.1109/LCA.2015.2434872.

[25] V. Seshadri, Y. Kim, C. Fallin, D. Lee, R. Ausavarungnirun, G. Pekhimenko, Y. Luo, O. Mutlu, P.B. Gibbons, M.A. Kozuch, T.C. Mowry, RowClone: fast and energy-efficient in-DRAM bulk data copy and initialization, in: Proceedings of the 46th Annual IEEE/ACM International Symposium on Microarchitecture, MICRO-46, Davis, CA, ACM, New York, NY, ISBN 978-1-4503-2638-4, 2013, pp. 185–197, http://dx.doi.org/10.1145/2540708.2540725.

[26] K. Hsieh, S. Khan, N. Vijaykumar, K.K. Chang, A. Boroumand, S. Ghose, O. Mutlu, Accelerating pointer chasing in 3D-stacked memory: challenges, mechanisms, evaluation, in: ICCD, 2016.

[27] M. Hashemi, O. Mutlu, Y.N. Patt, Continuous runahead: transparent hardware acceleration for memory intensive workloads, in: MICRO, 2016.

[28] M. Hashemi, Khubaib, E. Ebrahimi, O. Mutlu, Y.N. Patt, Accelerating dependent cache misses with an enhanced memory controller, in: ISCA, 2016.

[29] JEDEC, DDR4 SDRAM Standard. JEDEC (online publication), 2013, http://www.jedec.org/standards-documents/docs/jesd79-4a.

[30] Y. Kang, W. Huang, S.-M. Yoo, D. Keen, Z. Ge, V. Lam, P. Pattnaik, J. Torrellas, FlexRAM: toward an advanced intelligent memory system, in: Proceedings of the 1999 IEEE International Conference on Computer Design, ICCD '99, IEEE Computer Society, Washington, DC, ISBN 0-7695-0406-X, 1999, p. 192, http://dl.acm.org/citation.cfm?id=846215.846721.

[31] B.B. Fraguela, J. Renau, P. Feautrier, D. Padua, J. Torrellas, Programming the FlexRAM parallel intelligent memory system, in: Proceedings of the Ninth ACM SIGPLAN Symposium on Principles and Practice of Parallel Programming, PPoPP '03, San Diego, CA, USA, ACM, New York, NY, ISBN 1-58113-588-2, 2003, pp. 49–60, http://dx.doi.org/10.1145/781498.781505.

[32] J. Draper, J. Chame, M. Hall, C. Steele, T. Barrett, J. LaCoss, J. Granacki, J. Shin, C. Chen, C.W. Kang, I. Kim, G. Daglikoca, The architecture of the DIVA processing-in-memory chip, in: Proceedings of the 16th International Conference on Supercomputing, ICS '02, New York, NY, USA, ACM, New York, NY, ISBN 1-58113-483-5, 2002, pp. 14–25, http://dx.doi.org/10.1145/514191.514197.

[33] G.H. Loh, 3D-stacked memory architectures for multi-core processors, in: Proceedings of the 35th Annual International Symposium on Computer Architecture, ISCA '08, IEEE Computer Society, Washington, DC, ISBN 978-0-7695-3174-8, 2008, pp. 453–464, http://dx.doi.org/10.1109/ISCA.2008.15.

[34] J. Jeddeloh, B. Keeth, Hybrid memory cube: new DRAM architecture increases density and performance, in: Symposium on VLSI Technology, VLSIT, ISSN 0743-1562, 2012, pp. 87–88, http://dx.doi.org/10.1109/VLSIT.2012.6242474.

[35] High Bandwidth Memory DRAM, AMD, 2015, http://www.jedec.org/standards-documents/docs/jesd235

[36] A. Patnaik, X. Tang, A. Jog, O. Kayiran, A.K. Mishra, M.T. Kandemir, O. Mutlu, C.R. Das, Scheduling techniques for GPU architectures with processing-in-memory capabilities, in: PACT, 2016.

[37] A. Boroumand, S. Ghose, B. Lucia, K. Hsieh, K. Malladi, H. Zheng, O. Mutlu, LazyPIM: an efficient cache coherence mechanism for processing–in-memory, IEEE Comput. Archit. Lett. PP (99) (2016) 1, ISSN 1556-6056, http://dx.doi.org/10.1109/LCA.2016.2577557.

[38] S. Li, C. Xu, Q. Zou, J. Zhao, Y. Lu, Y. Xie, Pinatubo: a processing–in–memory architecture for bulk bitwise operations in emerging non-volatile memories, in: Proceedings of the 53rd Annual Design Automation Conference, ACM, 2016, pp. 173.

[39] B.C. Lee, E. Ipek, O. Mutlu, D. Burger, Architecting phase change memory as a scalable DRAM alternative, in: Proceedings of the 36th Annual International Symposium on Computer Architecture, ISCA '09, Austin, TX, USA, ACM, New York, NY, ISBN 978-1-60558-526-0, 2009, pp. 2–13, http://dx.doi.org/10.1145/1555754.1555758.

[40] M.K. Qureshi, V. Srinivasan, J.A. Rivers, Scalable high performance main memory system using phase-change memory technology, in: Proceedings of the 36th Annual International Symposium on Computer Architecture, ISCA '09, Austin, TX, USA, ACM, New York, NY, ISBN 978-1-60558-526-0, 2009, pp. 24–33, http://dx.doi.org/10.1145/1555754.1555760.

[41] H.S.P. Wong, S. Raoux, S. Kim, J. Liang, J.P. Reifenberg, B. Rajendran, M. Asheghi, K.E. Goodson, Phase change memory, Proc. IEEE 98 (12) (2010) 2201–2227, ISSN 0018-9219, http://dx.doi.org/10.1109/JPROC.2010.2070050.

[42] P. Zhou, B. Zhao, J. Yang, Y. Zhang, A Durable and Energy Efficient Main Memory Using Phase Change Memory Technology, in: Proceedings of the 36th Annual International Symposium on Computer Architecture, ISCA '09, Austin, TX, USA, ACM, New York, NY, ISBN 978-1-60558-526-0, 2009, pp. 14–23, http://dx.doi.org/10.1145/1555754.1555759.

[43] B.C. Lee, P. Zhou, J. Yang, Y. Zhang, B. Zhao, E. Ipek, O. Mutlu, D. Burger, Phase-change technology and the future of main memory, IEEE Micro 30 (1) (2010) 143–143, ISSN 0272-1732, http://dx.doi.org/10.1109/MM.2010.24.

[44] S. Raoux, G.W. Burr, M.J. Breitwisch, C.T. Rettner, Y.-C. Chen, R.M. Shelby, M. Salinga, D. Krebs, S.-H. Chen, H.-L. Lung, C.H. Lam, Phase-change random access memory: a scalable technology, IBM J. Res. Dev. 52 (4) (2008) 465–479, ISSN 0018-8646, http://dx.doi.org/10.1147/rd.524.0465.

[45] B.C. Lee, E. Ipek, O. Mutlu, D. Burger, Phase change memory architecture and the quest for scalability, Commun. ACM 53 (7) (2010) 99–106, ISSN 0001-0782, http://dx.doi.org/10.1145/1785414.1785441.

[46] M. Kang, M.-S. Keel, N.R. Shanbhag, S. Eilert, K. Curewitz, An energy-efficient VLSI architecture for pattern recognition via deep embedding of computation in SRAM, in: 2014 IEEE International Conference on Acoustics, Speech and Signal Processing (ICASSP), IEEE, 2014, pp. 8326–8330.

[47] A. Shafiee, A. Nag, N. Muralimanohar, R. Balasubramanian, J.P. Strachan, M. Hu, R.S. Williams, V. Srikumar, ISAAC: a convolutional neural network accelerator with in-situ analog arithmetic in crossbars, in: Proc. ISCA, 2016.

[48] Y. Kim, V. Seshadri, D. Lee, J. Liu, O. Mutlu, A case for exploiting subarray-level parallelism (SALP) in DRAM, in: Proceedings of the 39th Annual International Symposium on Computer Architecture, ISCA '12, Portland, Oregon, IEEE Computer Society, Washington, DC, ISBN 978-1-4503-1642-2, 2012, pp. 368–379, http://dl.acm.org/citation.cfm?id=2337159.2337202.

[49] D. Lee, Y. Kim, V. Seshadri, J. Liu, L. Subramanian, O. Mutlu, Tiered-latency DRAM: a low latency and low cost DRAM architecture, in: Proceedings of the 2013 IEEE 19th International Symposium on High Performance Computer Architecture (HPCA), HPCA '13, IEEE Computer Society, Washington, DC, ISBN 978-1-4673-5585-8, 2013, pp. 615–626, http://dx.doi.org/10.1109/HPCA.2013.6522354.

[50] D. Lee, Y. Kim, G. Pekhimenko, S.M. Khan, V. Seshadri, K.K.-W. Chang, O. Mutlu, Adaptive-latency DRAM: optimizing DRAM timing for the common-case, in: HPCA, IEEE, 2015, pp. 489–501, http://dx.doi.org/10.1109/HPCA.2015.7056057.

[51] V. Seshadri, T. Mullins, A. Boroumand, O. Mutlu, P.B. Gibbons, M.A. Kozuch, T.C. Mowry, Gather-scatter DRAM: in-DRAM address translation to improve the spatial locality of non-unit strided accesses, in: Proceedings of the 48th International

Symposium on Microarchitecture, MICRO-48, Waikiki, Hawaii, ACM, New York, NY, ISBN 978-1-4503-4034-2, 2015, pp. 267–280, http://dx.doi.org/10.1145/2830772.2830820.

[52] K.K.-W. Chang, D. Lee, Z. Chishti, A.R. Alameldeen, C. Wilkerson, Y. Kim, O. Mutlu, Improving DRAM performance by parallelizing refreshes with accesses, in: 2014 IEEE 20th International Symposium on High Performance Computer Architecture (HPCA), IEEE, 2014, pp. 356–367.

[53] Y. Kim, W. Yang, O. Mutlu, Ramulator: a fast and extensible DRAM simulator, IEEE Comput. Archit. Lett. 15 (1) (2016) 45–49, ISSN 1556-6056, http://dx.doi.org/10.1109/LCA.2015.2414456.

[54] J. Liu, B. Jaiyen, Y. Kim, C. Wilkerson, O. Mutlu, An Experimental Study of Data Retention Behavior in Modern DRAM Devices: Implications for Retention Time Profiling Mechanisms, in: Proceedings of the 40th Annual International Symposium on Computer Architecture, ISCA '13, Tel-Aviv, Israel, ACM, New York, NY, ISBN 978-1-4503-2079-5, 2013, pp. 60–71, http://dx.doi.org/10.1145/2485922.2485928.

[55] S.M. Khan, D. Lee, O. Mutlu, PARBOR: an efficient system-level technique to detect data-dependent failures in DRAM, in: DSN, 2016.

[56] K.K. Chang, A. Kashyap, H. Hassan, S. Ghose, K. Hsieh, D. Lee, T. Li, G. Pekhimenko, S. Khan, O. Mutlu, Understanding latency variation in modern DRAM chips: experimental characterization, analysis, and optimization, in: Sigmetrics, 2016.

[57] S. Khan, D. Lee, Y. Kim, A.R. Alameldeen, C. Wilkerson, O. Mutlu, The efficacy of error mitigation techniques for DRAM retention failures: a comparative experimental study, in: The 2014 ACM International Conference on Measurement and Modeling of Computer Systems, SIGMETRICS '14, Austin, Texas, USA, ACM, New York, NY, ISBN 978-1-4503-2789-3, 2014, pp. 519–532, http://dx.doi.org/10.1145/2591971.2592000.

[58] J. Liu, B. Jaiyen, R. Veras, O. Mutlu, RAIDR: retention-aware intelligent DRAM refresh, in: Proceedings of the 39th Annual International Symposium on Computer Architecture, ISCA '12, Portland, Oregon, IEEE Computer Society, Washington, DC, ISBN 978-1-4503-1642-2, 2012, pp. 1–12, http://dl.acm.org/citation.cfm?id=2337159.2337161.

[59] H. Hassan, G. Pekhimenko, N. Vijaykumar, V. Seshadri, D. Lee, O. Ergin, O. Mutlu, ChargeCache: reducing DRAM latency by exploiting row access locality, in: HPCA, 2016.

[60] M.K. Qureshi, D.H. Kim, S. Khan, P.J. Nair, O. Mutlu, AVATAR: a variable-retention-time (VRT) aware refresh for DRAM systems, in: 2015 45th Annual IEEE/IFIP International Conference on Dependable Systems and Networks, ISSN 1530-2015, pp. 427–437, http://dx.doi.org/10.1109/DSN.2015.58.

[61] Rambus, DRAM Power Model. 2010, https://www.rambus.com/energy/.

[62] A.N. Udipi, N. Muralimanohar, N. Chatterjee, R. Balasubramanian, A. Davis, N.P. Jouppi, Rethinking DRAM design and organization for energy-constrained multi-cores, in: Proceedings of the 37th Annual International Symposium on Computer Architecture, ISCA '10, Saint-Malo, France, ACM, New York, NY, ISBN 978-1-4503-0053-7, 2010, pp. 175–186, http://dx.doi.org/10.1145/1815961.1815983.

[63] T. Zhang, K. Chen, C. Xu, G. Sun, T. Wang, Y. Xie, Half-DRAM: a high-bandwidth and low-power DRAM architecture from the rethinking of fine-grained activation, in: Proceeding of the 41st Annual International Symposium on Computer Architecture, ISCA '14, Minneapolis, Minnesota, USA, IEEE Press, Piscataway, NJ, ISBN 978-1-4799-4394-4, 2014, pp. 349–360, http://dl.acm.org/citation.cfm?id=2665671.2665724.

[64] JEDEC, DDR3 SDRAM, JESD79-3F. 2012, https://www.jedec.org/sites/default/files/docs/JESD79-3F.pdf.

[65] X. Jiang, Y. Solihin, L. Zhao, R. Iyer, Architecture support for improving bulk memory copying and initialization performance, in: Proceedings of the 2009 18th International Conference on Parallel Architectures and Compilation Techniques (PACT), IEEE Computer Society, Washington, DC, USA, ISBN 978-0-7695-3771-9, 2009, pp. 169–180, http://dx.doi.org/10.1109/PACT.2009.31.

[66] X. Yang, S.M. Blackburn, D. Frampton, J.B. Sartor, K.S. McKinley, Why nothing matters: the impact of zeroing, in: Proceedings of the 2011 ACM International Conference on Object Oriented Programming Systems Languages and Applications (OOPSLA), Portland, Oregon, USA, ACM, New York, NY, ISBN 978-1-4503-0940-0, 2011, pp. 307–324, http://dx.doi.org/10.1145/2048066.2048092.

[67] C.-Y. Chan, Y.E. Ioannidis, Bitmap index design and evaluation, in: Proceedings of the 1998 ACM SIGMOD International Conference on Management of Data, SIGMOD '98, Seattle, Washington, USA, ACM, New York, NY, ISBN 0-89791-995-5, 1998, pp. 355–366, http://dx.doi.org/10.1145/276304.276336.

[68] E. O'Neil, P. O'Neil, K. Wu, Bitmap index design choices and their performance implications, in: Proceedings of the 11th International Database Engineering and Applications Symposium, IDEAS '07, IEEE Computer Society, Washington, DC, USA, ISBN 0-7695-2947-X, 2007, pp. 72–84, http://dx.doi.org/10.1109/IDEAS.2007.19.

[69] FastBit: An Efficient Compressed Bitmap Index Technology, https://sdm.lbl.gov/fastbit/ (Accessed December 2014).

[70] K. Wu, E.J. Otoo, A. Shoshani, Compressing bitmap indexes for faster search operations, in: Proceedings of the 14th International Conference on Scientific and Statistical Database Management, SSDBM '02, IEEE Computer Society, Washington, DC, ISBN 0-7695-1632-7, 2002, pp. 99–108, http://dx.doi.org/10.1109/SSDM.2002.1029710.

[71] Redis—Bitmaps. http://redis.io/topics/data-types-intro#bitmaps (Accessed August 2015).

[72] B. Keeth, R.J. Baker, B. Johnson, F. Lin, DRAM Circuit Design: Fundamental and High-Speed Topics, second ed., Wiley-IEEE Press, Hoboken, NJ, USA, 2007, ISBN 0470184752, 9780470184752.

[73] Intel, Intel 64 and IA-32 Architectures Optimization Reference Manual. 2012, https://www.intel.com/content/dam/www/public/us/en/documents/manuals/64-ia-32-architectures-optimization-manual.pdf.

[74] I.B.M. Corporation, Enterprise Systems Architecture/390 Principles of Operation, 2001, https://www-05.ibm.com/e-business/linkweb/publications/servlet/pbi.wss?CTY=US&FNC=SRX&PBL=SA22-7201-08.

[75] Intel, Intel 64 and IA-32 Architectures Software Developer's Manual, vol. 3A, 2012, p. 12 (Chapter 11). https://www.intel.com/content/dam/www/public/us/en/documents/manuals/64-ia-32-architectures-optimization-manual.pdf.

[76] JEDEC, Standard No. 21-C. Annex K: Serial Presence Detect (SPD) for DDR3 SDRAM Modules, 2011, https://www.jedec.org/sites/default/files/docs/4_01_02_11R21A.pdf.

[77] M. Horiguchi, K. Itoh, Nanoscale Memory Repair, Springer, New York City, NY, USA, 2011.

[78] F.A. Ware, C. Hampel, Improving power and data efficiency with threaded memory modules, in: ICCD, 2006.

[79] H. Zheng, J. Lin, Z. Zhang, E. Gorbatov, H. David, Z. Zhu, Mini-rank: adaptive DRAM architecture for improving memory power efficiency, in: Proceedings of the 41st Annual IEEE/ACM International Symposium on Microarchitecture, MICRO, 2008.

[80] V. Seshadri, O. Mutlu, M.A. Kozuch, T.C. Mowry, The evicted-address filter: a
      unified mechanism to address both cache pollution and thrashing, in: Proceedings
      of the 21st International Conference on Parallel Architectures and Compilation
      Techniques, PACT '12, Minneapolis, Minnesota, USA, ACM, New York, NY,
      ISBN 978-1-4503-1182-3, 2012, pp. 355–366, http://dx.doi.org/10.1145/
      2370816.2370868.
[81] V. Seshadri, S. Yedkar, H. Xin, O. Mutlu, P.B. Gibbons, M.A. Kozuch, T.C. Mowry,
      Mitigating prefetcher-caused pollution using informed caching policies for prefetched
      blocks, ACM Trans. Archit. Code Optim. 11 (4) (2015) 51:1–51:22, ISSN 1544-3566,
      http://dx.doi.org/10.1145/2677956.
[82] Safari, Memsim. http://safari.ece.cmu.edu/tools.html, 2012.
[83] Github, Ramulator Source Code. https://github.com/CMU-SAFARI/ramulator,
      2015.
[84] Memcached: a high performance, distributed memory object caching system. http://
      memcached.org (Accessed October 2012).
[85] MySQL: an open source database. http://www.mysql.com (Accessed October 2012).
[86] Standard Performance Evaluation Corporation, SPEC CPU2006 Benchmark Suite.
      www.spec.org/cpu2006, 2006.
[87] Bochs IA-32 Emulator Project. http://bochs.sourceforge.net/ (Accessed October
      2012).
[88] Wind River Simics Full System Simulation. http://www.windriver.com/products/
      simics/ (Accessed January 2013).
[89] S. Eyerman, L. Eeckhout, System-level performance metrics for multiprogram work-
      loads, IEEE Micro 28 (3) (2008) 42–53, ISSN 0272-1732, http://dx.doi.org/10.1109/
      MM.2008.44.
[90] A. Snavely, D.M. Tullsen, Symbiotic jobscheduling for a simultaneous multithrea-
      ded processor, in: Proceedings of the Ninth International Conference on Architectural
      Support for Programming Languages and Operating Systems, ASPLOS IX,
      Cambridge, Massachusetts, USA, ACM, New York, NY, ISBN 1-58113-317-0,
      2000, pp. 234–244, http://dx.doi.org/10.1145/378993.379244.
[91] V. Seshadri, A. Bhowmick, O. Mutlu, P.B. Gibbons, M.A. Kozuch, T.C. Mowry,
      The dirty-block index, in: Proceeding of the 41st Annual International Symposium
      on Computer Architecture, ISCA '14, Minneapolis, Minnesota, USA, IEEE Press,
      Piscataway, NJ, ISBN 978-1-4799-4394-4, 2014, pp. 157–168, http://dl.acm.org/
      citation.cfm?id=2665671.2665697.
[92] K.K.-W. Chang, R. Ausavarungnirun, C. Fallin, O. Mutlu, HAT: heterogeneous
      adaptive throttling for on-chip networks, in: Proceedings of the 2012 IEEE 24th Inter-
      national Symposium on Computer Architecture and High Performance Computing,
      SBAC-PAD '12, IEEE Computer Society, Washington, DC, ISBN 978-0-7695-
      4907-1, 2012, pp. 9–18, http://dx.doi.org/10.1109/SBAC-PAD.2012.44.
[93] L. Subramanian, D. Lee, V. Seshadri, H. Rastogi, O. Mutlu, The blacklisting memory
      scheduler: achieving high performance and fairness at low cost, in: 32nd IEEE Inter-
      national Conference on Computer Design (ICCD), 2014, 2014.
[94] L. Subramanian, D. Lee, V. Seshadri, H. Rastogi, O. Mutlu, BLISS: balancing perfor-
      mance, fairness and complexity in memory access scheduling, in: 32nd IEEE Interna-
      tional Conference on Computer Design (ICCD), 2014, IEEE Trans. Parallel Distrib.
      Syst. 27 (10) (2016), 3071–3087.
[95] H. Yoon, J. Meza, R. Ausavarungnirun, R. Harding, O. Mutlu, Row Buffer Locality
      Aware Caching Policies for Hybrid Memories, in: Proceedings of the 2012 IEEE 30th
      International Conference on Computer Design (ICCD 2012), ICCD '12, IEEE
      Computer Society, Washington, DC, ISBN 978-1-4673-3051-0, 2012,
      pp. 337–344, http://dx.doi.org/10.1109/ICCD.2012.6378661.

[96] J. Meza, J. Li, O. Mutlu, A case for small row buffers in non-volatile main memories, in: Proceedings of the 2012 IEEE 30th International Conference on Computer Design (ICCD 2012), ICCD '12, IEEE Computer Society, Washington, DC, ISBN 978-1-4673-3051-0, 2012, pp. 484–485, http://dx.doi.org/10.1109/ICCD.2012.6378685.

[97] H. Yoon, J. Meza, N. Muralimanohar, N.P. Jouppi, O. Mutlu, Efficient data mapping and buffering techniques for multilevel cell phase-change memories, ACM Trans. Archit. Code Optim. 11 (4) (2014) 40:1–40:25, ISSN 1544-3566, http://dx.doi.org/10.1145/2669365.

[98] E. Ebrahimi, O. Mutlu, C.J. Lee, Y.N. Patt, Coordinated control of multiple prefetchers in multi-core systems, in: Proceedings of the 42nd Annual IEEE/ACM International Symposium on Microarchitecture, MICRO 42, New York, NY, ACM, New York, NY, ISBN 978-1-60558-798-1, 2009, pp. 316–326, http://dx.doi.org/10.1145/1669112.1669154.

[99] R.F. Sauers, C.P. Ruemmler, P.S. Weygant, Memory bottlenecks, in: HP-UX 11i Tuning and Performance, Prentice Hall, Upper Saddle River, NJ, 2004 (Chapter 8).

[100] L. Subramanian, V. Seshadri, Y. Kim, B. Jaiyen, O. Mutlu, MISE: providing performance predictability and improving fairness in shared main memory systems, in: IEEE 19th International Symposium on High Performance Computer Architecture, ISSN 1530-0897, 2013, pp. 639–650, http://dx.doi.org/10.1109/HPCA.2013.6522356.

[101] L. Subramanian, V. Seshadri, A. Ghosh, S. Khan, O. Mutlu, The application slowdown model: quantifying and controlling the impact of inter-application interference at shared caches and main memory, in: Proceedings of the 48th International Symposium on Microarchitecture, MICRO-48, Waikiki, Hawaii, ACM, New York, NY, ISBN 978-1-4503-4034-2, 2015, pp. 62–75, http://dx.doi.org/10.1145/2830772.2830803.

[102] O. Mutlu, L. Subramanian, Research problems and opportunities in memory systems. Supercomput. Front. Innov. 1 (3) (2014) 19–55.

[103] O. Mutlu, Memory scaling: a systems architecture perspective, in: IMW, 2014.

[104] Y. Kim, R. Daly, J. Kim, C. Fallin, J.H. Lee, D. Lee, C. Wilkerson, K. Lai, O. Mutlu, Flipping bits in memory without accessing them: an experimental study of DRAM disturbance errors, in: Proceeding of the 41st Annual International Symposium on Computer Architecture, ISCA '14, Minneapolis, Minnesota, USA, IEEE Press, Piscataway, NJ, ISBN 978-1-4799-4394-4, 2014, pp. 361–372, http://dl.acm.org/citation.cfm?id=2665671.2665726.

[105] O. Mutlu, Efficient Runahead Execution Processors (Ph.D. thesis), Austin, TX, 2006, aAI3263366.

[106] O. Mutlu, J. Stark, C. Wilkerson, Y.N. Patt, Runahead execution: an alternative to very large instruction windows for out-of-order processors, in: Proceedings of the 9th International Symposium on High-Performance Computer Architecture, HPCA '03, IEEE Computer Society, Washington, DC, USA, ISBN 0-7695-1871-0, 2003, p. 129, http://dl.acm.org/citation.cfm?id=822080.822823.

[107] O. Mutlu, T. Moscibroda, Parallelism-aware batch scheduling: enhancing both performance and fairness of shared DRAM systems, in: Proceedings of the 35th Annual International Symposium on Computer Architecture, ISCA '08, IEEE Computer Society, Washington, DC, ISBN 978-0-7695-3174-8, 2008, pp. 63–74, http://dx.doi.org/10.1109/ISCA.2008.7.

[108] C.J. Lee, V. Narasiman, O. Mutlu, Y.N. Patt, Improving memory bank-level parallelism in the presence of prefetching, in: Proceedings of the 42nd Annual IEEE/ACM International Symposium on Microarchitecture, MICRO 42, New York, NY, ACM, New York, NY, ISBN 978-1-60558-798-1, 2009, pp. 327–336, http://dx.doi.org/10.1145/1669112.1669155.

[109] A. Glew, MLP yes! ILP no, In Wild and Crazy Ideas Session, 8th International Conference on Architectural Support for Programming Languages and Operating Systems, October 1998.

[110] R. Das, O. Mutlu, T. Moscibroda, C.R. Das, Application-aware prioritization mechanisms for on-chip networks, in: Proceedings of the 42Nd Annual IEEE/ACM International Symposium on Microarchitecture, MICRO 42, New York, NY, ACM, New York, NY, ISBN 978-1-60558-798-1, 2009, pp. 280–291, http://dx.doi.org/10.1145/1669112.1669150.

[111] Y. Kim, M. Papamichael, O. Mutlu, M. Harchol-Balter, Thread cluster memory scheduling: exploiting differences in memory access behavior, in: Proceedings of the 2010 43rd Annual IEEE/ACM International Symposium on Microarchitecture, MICRO '43, IEEE Computer Society, Washington, DC, USA, ISBN 978-0-7695-4299-7, 2010, pp. 65–76, http://dx.doi.org/10.1109/MICRO.2010.51.

[112] Y. Kim, D. Han, O. Mutlu, M. Harchol-Balter, ATLAS: a scalable and high-performance scheduling algorithm for multiple memory controllers, in: IEEE 16th International Symposium on High Performance Computer Architecture, ISSN 1530-0897, 2010, pp. 1–12, http://dx.doi.org/10.1109/HPCA.2010.5416658.

[113] S.P. Muralidhara, L. Subramanian, O. Mutlu, M. Kandemir, T. Moscibroda, Reducing memory interference in multicore systems via application-aware memory channel partitioning, in: Proceedings of the 44th Annual IEEE/ACM International Symposium on Microarchitecture, MICRO-44, Porto Alegre, Brazil, ACM, New York, NY, ISBN 978-1-4503-1053-6, 2011, pp. 374–385, http://dx.doi.org/10.1145/2155620.2155664.

[114] E. Ebrahimi, C.J. Lee, O. Mutlu, Y.N. Patt, Fairness via source throttling: a configurable and high-performance fairness substrate for multi-core memory systems, in: Proceedings of the Fifteenth Edition of ASPLOS on Architectural Support for Programming Languages and Operating Systems, ASPLOS XV, Pittsburgh, Pennsylvania, USA, ACM, New York, NY, ISBN 978-1-60558-839-1, 2010, pp. 335–346, http://dx.doi.org/10.1145/1736020.1736058.

[115] R. Ausavarungnirun, K.K.-W. Chang, L. Subramanian, G.H. Loh, O. Mutlu, Staged memory scheduling: achieving high performance and scalability in heterogeneous systems, in: Proceedings of the 39th Annual International Symposium on Computer Architecture, ISCA '12, Portland, Oregon, IEEE Computer Society, Washington, DC, ISBN 978-1-4503-1642-2, 2012, pp. 416–427, http://dl.acm.org/citation.cfm?id=2337159.2337207.

[116] H. Vandierendonck, A. Seznec, Fairness metrics for multi-threaded processors, IEEE Comput. Archit. Lett. 10 (1) (2011) 4–7, ISSN 1556-6056, http://dx.doi.org/10.1109/L-CA.2011.1.

[117] R. Das, R. Ausavarungnirun, O. Mutlu, A. Kumar, M. Azimi, Application-to-core mapping policies to reduce memory interference in multi-core systems, in: Proceedings of the 21st International Conference on Parallel Architectures and Compilation Techniques, PACT '12, Minneapolis, Minnesota, USA, ACM, New York, NY, ISBN 978-1-4503-1182-3, 2012, pp. 455–456, http://dx.doi.org/10.1145/2370816.2370893.

[118] J. Zhao, O. Mutlu, Y. Xie, FIRM: fair and high-performance memory control for persistent memory systems, in: Proceedings of the 47th Annual IEEE/ACM International Symposium on Microarchitecture, MICRO-47, Cambridge, United Kingdom, IEEE Computer Society, Washington, DC, ISBN 978-1-4799-6998-2, 2014, pp. 153–165, http://dx.doi.org/10.1109/MICRO.2014.47.

[119] H. Usui, L. Subramanian, K.K.-W. Chang, O. Mutlu, DASH: deadline-aware high-performance memory scheduler for heterogeneous systems with hardware accelerators, ACM Trans. Archit. Code Optim. 12 (4) (2016) 65:1–65:28, ISSN 1544-3566, http://dx.doi.org/10.1145/2847255.

[120] H. Kim, D. de Niz, B. Andersson, M. Klein, O. Mutlu, R. Rajkumar, Bounding and reducing memory interference in COTS-based multi-core systems, Real-Time Syst. 52 (3) (2016) 356–395, ISSN 1573-1383, http://dx.doi.org/10.1007/s11241-016-9248-1.

[121] R. Das, O. Mutlu, T. Moscibroda, C.R. Das, Aérgia: exploiting packet latency slack in on-chip networks, in: Proceedings of the 37th Annual International Symposium on Computer Architecture, ISCA '10, Saint-Malo, France, ACM, New York, NY, ISBN 978-1-4503-0053-7, 2010, pp. 106–116, http://dx.doi.org/10.1145/1815961.1815976.

[122] E. Ebrahimi, C.J. Lee, O. Mutlu, Y.N. Patt, Prefetch-aware shared resource management for multi-core systems, in: Proceedings of the 38th Annual International Symposium on Computer Architecture, ISCA '11, San Jose, California, USA, ACM, New York, NY, ISBN 978-1-4503-0472-6, 2011, pp. 141–152, http://dx.doi.org/10.1145/2000064.2000081.

[123] R. Das, R. Ausavarungnirun, O. Mutlu, A. Kumar, M. Azimi, Application-to-core mapping policies to reduce memory interference in multi-core systems, in: HPCA, 2012.

[124] H. Kim, D. de Niz, B. Andersson, M. Klein, O. Mutlu, R. Rajkumar, Bounding and reducing memory interference in COTS-based multi-core systems, in: RTAS, 2014.

[125] K. Luo, J. Gummaraju, M. Franklin, Balancing thoughput and fairness in SMT processors, in: IEEE International Symposium on Performance Analysis of Systems and Software, 2001, 2001, pp. 164–171, http://dx.doi.org/10.1109/ISPASS.2001.990695.

[126] S. Srinath, O. Mutlu, H. Kim, Y.N. Patt, Feedback directed prefetching: improving the performance and bandwidth-efficiency of hardware prefetchers, in: Proceedings of the 2007 IEEE 13th International Symposium on High Performance Computer Architecture, HPCA '07, IEEE Computer Society, Washington, DC, USA, ISBN 1-4244-0804-0, 2007, pp. 63–74, http://dx.doi.org/10.1109/HPCA.2007.346185.

[127] L. Zhao, L.N. Bhuyan, R. Iyer, S. Makineni, D. Newell, Hardware support for accelerating data movement in server platform, IEEE Trans. Comput. 56 (6) (2007) 740–753, ISSN 0018-9340, http://dx.doi.org/10.1109/TC.2007.1036.

[128] D.P. Bovet, M. Cesati, Understanding the Linux Kernel, in: O'Reilly Media, 2005, p. 388.

[129] M.E. Russinovich, D.A. Solomon, A. Ionescu, Windows Internals, Microsoft Press, Redmond, WA, USA, 2009, 701 pp.

[130] A. Singh, Mac OS X Internals: A Systems Approach, Addison-Wesley Professional, Boston, MA, 2006.

[131] J. Chow, B. Pfaff, T. Garfinkel, M. Rosenblum, Shredding your garbage: reducing data lifetime through secure deallocation, in: Proceedings of the 14th Conference on USENIX Security Symposium, SSYM'05, Baltimore, MD, vol. 14, USENIX Association, Berkeley, CA, 2005, pp. 22–22, http://dl.acm.org/citation.cfm?id=1251398.1251420.

[132] A.M. Dunn, M.Z. Lee, S. Jana, S. Kim, M. Silberstein, Y. Xu, V. Shmatikov, E. Witchel, Eternal sunshine of the spotless machine: protecting privacy with ephemeral channels, in: Proceedings of the 10th USENIX Conference on Operating Systems Design and Implementation, OSDI'12, Hollywood, CA, USA, USENIX Association, Berkeley, CA, USA, ISBN 978-1-931971-96-6, 2012, pp. 61–75, http://dl.acm.org/citation.cfm?id=2387880.2387887.

[133] J.A. Halderman, S.D. Schoen, N. Heninger, W. Clarkson, W. Paul, J.A. Calandrino, A.J. Feldman, J. Appelbaum, E.W. Felten, Lest we remember: cold-boot attacks on encryption keys, Commun. ACM 52 (5) (2009) 91–98, ISSN 0001-0782, http://dx.doi.org/10.1145/1506409.1506429.

[134] K. Harrison, S. Xu, Protecting cryptographic keys from memory disclosure attacks, in: 37th Annual IEEE/IFIP International Conference on Dependable Systems and Networks, 2007, pp. 137–143, http://dx.doi.org/10.1109/DSN.2007.77.

[135] J. Bent, G. Gibson, G. Grider, B. McClelland, P. Nowoczynski, J. Nunez, M. Polte, M. Wingate, PLFS: a checkpoint filesystem for parallel applications, in: Proceedings of the Conference on High Performance Computing Networking, Storage and Analysis, SC '09, Portland, Oregon, ACM, New York, NY, ISBN 978-1-60558-744-8, 2009, pp. 21:1–21:12, http://dx.doi.org/10.1145/1654059.1654081.

[136] S.M. Srinivasan, S. Kandula, C.R. Andrews, Y. Zhou, Flashback: a lightweight extension for rollback and deterministic replay for software debugging, in: Proceedings of the Annual Conference on USENIX Annual Technical Conference, ATEC '04, Boston, MA, USENIX Association, Berkeley, CA, 2004, pp. 3, http://dl.acm.org/citation.cfm?id=1247415.1247418.

[137] K. Constantinides, O. Mutlu, T. Austin, V. Bertacco, Software-based online detection of hardware defects mechanisms, architectural support, and evaluation, in: Proceedings of the 40th Annual IEEE/ACM International Symposium on Microarchitecture, MICRO 40, IEEE Computer Society, Washington, DC, ISBN 0-7695-3047-8, 2007, pp. 97–108, http://dx.doi.org/10.1109/MICRO.2007.39.

[138] K. Constantinides, O. Mutlu, T. Austin, Online Design Bug Detection: RTL Analysis, Flexible Mechanisms, and Evaluation, in: Proceedings of the 41st Annual IEEE/ACM International Symposium on Microarchitecture, MICRO 41, IEEE Computer Society, Washington, DC, ISBN 978-1-4244-2836-6, 2008, pp. 282–293, http://dx.doi.org/10.1109/MICRO.2008.4771798.

[139] K. Constantinides, O. Mutlu, T. Austin, V. Bertacco, A flexible software-based framework for online detection of hardware defects, IEEE Trans. Comput. 58 (8) (2009) 1063–1079, ISSN 0018-9340, http://doi.ieeecomputersociety.org/10.1109/TC.2009.52.

[140] J. Ren, J. Zhao, S. Khan, J. Choi, Y. Wu, O. Mutlu, ThyNVM: enabling software-transparent crash consistency in persistent memory systems, in: Proceedings of the 48th International Symposium on Microarchitecture, MICRO-48, Waikiki, Hawaii, ACM, New York, NY, ISBN 978-1-4503-4034-2, 2015, pp. 672–685, http://dx.doi.org/10.1145/2830772.2830802.

[141] F. Chang, G.A. Gibson, Automatic I/O hint generation through speculative execution, in: Proceedings of the Third Symposium on Operating Systems Design and Implementation, OSDI '99, New Orleans, Louisiana, USA, USENIX Association, Berkeley, CA, ISBN 1-880446-39-1, 1999, pp. 1–14, http://dl.acm.org/citation.cfm?id=296806.296807.

[142] B. Wester, P.M. Chen, J. Flinn, Operating system support for application-specific speculation, in: Proceedings of the Sixth Conference on Computer Systems, EuroSys '11, Salzburg, Austria, ACM, New York, NY, ISBN 978-1-4503-0634-8, 2011, pp. 229–242, http://dx.doi.org/10.1145/1966445.1966467.

[143] H.A. Lagar-Cavilla, J.A. Whitney, A.M. Scannell, P. Patchin, S.M. Rumble, E. de Lara, M. Brudno, M. Satyanarayanan, SnowFlock: rapid virtual machine cloning for cloud computing, in: Proceedings of the 4th ACM European Conference on Computer Systems, EuroSys '09, Nuremberg, Germany, ACM, New York, NY, ISBN 978-1-60558-482-9, 2009, pp. 1–12, http://dx.doi.org/10.1145/1519065.1519067.

[144] C.A. Waldspurger, Memory resource management in VMware ESX server, SIGOPS Oper. Syst. Rev. 36 (SI) (2002) 181–194, ISSN 0163-5980, http://dx.doi.org/10.1145/844128.844146.

[145] K. Sudan, N. Chatterjee, D. Nellans, M. Awasthi, R. Balasubramonian, A. Davis, Micro-pages: increasing DRAM efficiency with locality-aware data placement, in: Proceedings of the Fifteenth Edition of ASPLOS on Architectural Support for Programming Languages and Operating Systems, ASPLOS XV, Pittsburgh, Pennsylvania, USA, ACM, New York, NY, ISBN 978-1-60558-839-1, 2010, pp. 219–230, http://dx.doi.org/10.1145/1736020.1736045.

[146] T.B. Jablin, P. Prabhu, J.A. Jablin, N.P. Johnson, S.R. Beard, D.I. August, Automatic CPU-GPU communication management and optimization, in: Proceedings of the 32nd ACM SIGPLAN Conference on Programming Language Design and Implementation, PLDI '11, San Jose, California, USA, ACM, New York, NY, ISBN 978-1-4503-0663-8, 2011, pp. 142–151, http://dx.doi.org/10.1145/1993498.1993516.

[147] D. Lee, L. Subramanian, R. Ausavarungnirun, J. Choi, O. Mutlu, Decoupled Direct Memory Access: Isolating CPU and IO Traffic by Leveraging a Dual-Data-Port DRAM, in: Proceedings of the 2015 International Conference on Parallel Architecture and Compilation (PACT), PACT '15, IEEE Computer Society, Washington, DC, ISBN 978-1-4673-9524-3, 2015, pp. 174–187, http://dx.doi.org/10.1109/PACT.2015.51.

[148] The STAR Experiment. http://www.star.bnl.gov/ (Accessed December 2014).

[149] K.K. Chang, P.J. Nair, D. Lee, S. Ghose, M.K. Qureshi, O. Mutlu, Low-cost interlinked subarrays (LISA): enabling fast inter-subarray data movement in DRAM, in: HPCA, 2016.

ABOUT THE AUTHORS

Vivek Seshadri is a researcher at Microsoft Research India. His main interests are in developing new techniques to improve the performance and efficiency of computer systems. Prior to his current position, Vivek received his Bachelors in Computer Science and Engineering from Indian Institute of Technology, Madras in 2005, and his Ph.D. in Computer Science from Carnegie Mellon University in 2016. His Ph.D. thesis proposes new virtual memory and DRAM interfaces to enable highly efficient memory hierarchies.

Onur Mutlu is a professor of Computer Science at ETH Zurich. He is also a faculty member at Carnegie Mellon University, where he previously held the William D. and Nancy W. Strecker Early Career Professorship. His current broader research interests are in computer architecture, systems, and bioinformatics. He is especially interested in interactions across domains and between applications, system software, compilers, and microarchitecture, with a major current focus on memory and storage systems. He obtained his Ph.D. and MS in ECE from the University of Texas at Austin and BS degrees in Computer Engineering and Psychology from the University of Michigan, Ann Arbor. His industrial experience spans starting the Computer Architecture Group at Microsoft Research (2006–2009), and various product and research positions at Intel Corporation, Advanced Micro Devices, VMware, and Google. He received the inaugural IEEE Computer Society Young Computer Architect Award, the inaugural Intel Early Career Faculty Award, faculty partnership awards from various companies, and a healthy number of best paper or "Top Pick" paper recognitions at various computer systems and architecture venues. His computer architecture course lectures and materials are freely available on YouTube, and his research group makes software artifacts freely available online. For more information, please see his webpage at https://people.inf.ethz.ch/omutlu.

CHAPTER FIVE

A Novel Infrastructure for Synergistic Dataflow Research, Development, Education, and Deployment: The Maxeler AppGallery Project

Nemanja Trifunovic*, Boris Perovic†, Petar Trifunovic*, Zoran Babovic*, Ali R. Hurson‡
*School of Electrical Engineering, University of Belgrade, Belgrade, Serbia
†EPFL, Lausanne, Switzerland
‡Missouri University of Science and Technology, Rolla, MO, United States

Contents

Advances in Computers, Volume 106
ISSN 0065-2458
http://dx.doi.org/10.1016/bs.adcom.2017.04.005

Abstract

This chapter presents the essence and the details of a novel infrastructure that synergizes research, development, education, and deployment in the context of dataflow research. To make it clearer to fundamental scientists, the essence of the approach is explained by referencing the results of the work of four different Nobel laureates. To make it clearer to research community, crucial details are presented in the form of a manual. Till this point, the development of dataflow applications was based on tools inherited from the controlflow environment. We here describe a set of tools developed from scratch with dataflow specifics in mind. These tools are not only tuned to the dataflow environment, but they are also tuned to synergize with each other, for the best possible performance in minimal time, counting from the moment when new researchers enter the dataflow arena, until the moment when they are able to deliver a quality code for maximal speed performance and minimal energy consumption. The effectiveness of the presented synergetic approach was measured empirically, using a group of students in a dataflow course. The measured results clearly indicate the superiority of the proposed approach in the following five domains: time to design, time to program, time to build, time to test, and speedup ratio.

ABBREVIATIONS

APP Application
BRAM Block Random Access Memory
CA Content Administrator
CPU Central Processing Unit
DAPI DFE Creation Interface
DEV Developer
DFE Dataflow Engine
DSP Digital Signal Processor
FPGA Field-Programmable Gate Array
MAPI Multiple Action API
MAX-UP Maxeler University Program
NREG Non-Registered User
PI Principal Investigator
PI-A Principal Investigator Administrator
RAM Random Access Memory

REG Registered User
REST Representational State Transfer
SA Super Administrator
SAPI Single-DFE Single Action API
SLiC Simple Live CPU
UA User Administrator

1. INTRODUCTION

The tremendous growth of the amount of generated data from social networks, mobile networks, Internet-scale applications such as auctions, as well as in other domains such as the financial sector, science, the Internet of Things, and others raised new challenges regarding efficient processing and analysis of such Big Data. Researchers investigated different architectures for solving these Big Data problems, and over time software and hardware dataflow platforms established themselves as the most effective ones. In order to satisfy low-latency requirements posed by users, software dataflow solutions are characterized by a directed compute graph which consists of operators through which data flows. Spark [1] and Google Dataflow Platform [2] are the most popular platforms of this kind, which perform data processing typically in the cloud environment. However, hardware dataflow solutions are characterized by data-driven execution and they enable a low-latency parallel computing platform. Designed in the 1970s and 1980s, hardware dataflow solutions were not widely accepted until recently. The reasons for nonacceptance of early hardware dataflow systems varied, but in general the technology at the time could not efficiently support the requirements of such designs, especially the hardware resources necessary for handling data structures and communication [3]. Therefore, controlflow architectures implemented in standard microprocessor technology have dominated for over a few decades, both on supercomputing and personal computing markets. Consequently, software developers and scientists are not familiar with the process of programming dataflow systems, which is characterized by thinking-in-space, in contrast to the thinking-in-time approach, which is common for controlflow systems.

The major contribution of this chapter is that it introduces a set of fully developed and tested tools for development of dataflow applications, which are designed to meet the major challenges of dataflow supercomputing. These challenges include the slow pace at which even experienced controlflow programmers adopt the ways of thinking typical of dataflow, and the fact that programmers new to dataflow, even if experienced in controlflow programming, generate applications with speedups often much slower than the potentials of modern dataflow systems [4].

Introducing new ways of thinking and novel ideas may be challenging, since it is often met with resistance. Mahatma Gandhi summarized this problem very well in one of his famous quotes:

First they ignore you,
then they laugh at you,
then they fight you,
then you win.

There are many reasons why people resist to change their existing ways of thinking and their established ways of doing things in favor of new and often superior ways. Some of the reasons for initial resistance when new and novel ideas are introduced are:
- Fear of the unknown;
- Lack of competence;
- Psychological connection to the old ways and old ideas;
- Low trust in new ways and novel ideas;
- Fear that new ways are just a temporary fad;
- Difficulty to change existing routines;
- Need for effective training;
- Lack of awareness of possible cost benefits.

The AppGallery system was designed to address each one of the abovementioned eight issues. The stress was on visualizing the issues of importance in a way that makes AppGallery users fully aware of dataflow benefits: they can easily click to see details new to them, which eliminates the fear of unknown; they quickly reach the needed level of performance, so they start feeling competent, they get psychologically connected to the new ways of thinking, and they become trustful toward the new technology.

To minimize the resistance and to speed up the adoption process when introducing a new way of thinking or a novel idea, it is best to present it in an engaging way, within the scope of the existing infrastructure. The Maxeler AppGallery project and the Maxeler WebIDE project are an effort to expose the dataflow programming paradigm to a wide audience in an engaging and interesting way. The WebIDE project synergizes web-based design tools with operating system like environment, while the AppGallery project synergizes educational methodologies with educational needs in the process of acquisition of new technologies or new ways of thinking.

In addition, the Maxeler AppGallery project and the Maxeler WebIDE project are an effort meant to enhance the educational process at the School of Electrical Engineering at the University of Belgrade, in the field of Computer Engineering. The generation of the AppGallery system was eased by the fact that the work was done in an environment that had already been relatively sophisticated regarding the use of software-based tools for teaching and learning the advances in computer architecture and organization. The major references to the existing tools are: [5–14]. Ref. [5] describes an infrastructure for teaching intelligent systems. Ref. [6] reviews a wide plethora of systems that use multimedia effects to enhance the effectiveness of e-learning. Ref. [7] introduces a system for education in the domain of wireless sensor networks. Ref. [8] introduces a system for digital logic design and simulation. Ref. [9] describes an environment that supports a laboratory for computer architecture, and Ref. [10] evaluates a large variety of simulators that support teaching of computer architecture and organization. Ref. [11] describes a laboratory environment for computer-aided learning of computer architecture. Ref. [12] describes a web-based simulator of a pipelined processor used for educational purposes. Ref. [13] deals with a memory system for education. Finally, Ref. [14] introduces another web-based system for computer architecture and organization.

All of the aforementioned systems deal with issues related to teaching and learning of traditional concepts in the field of computer architecture and organization, namely, controlflow. However, the major question addressed in this research is how to create a framework for effective teaching and learning of a new paradigm in computer architecture and organization, namely, dataflow. This means that the tool should both support teaching and learning on one hand, and create motivation and eliminate fears of new concepts on the other hand.

The Maxeler AppGallery project was architected and designed specifically to address these two crucial sets of issues: enhancing teaching and learning capabilities (a) by closely coupling the selected dataflow tutorials and the web-based tool referred to as WebIDE, and (b) by closely coupling a gallery of best student achievements with explanatory powerpoint presentations that strengthen the motivation and eliminate the fear of the unknown.

This effort was inspired by issues related to the Routledge Complexity of Education [15]. The notions stressed in this research imply a synergy of mechanisms that enhance learning capabilities and learning motivation, and eliminate the fear of unknown. Notions of the Routledge Complexity of Education were closely followed during the process of architecting and designing the Maxeler AppGallery system.

Issues related to the impact of temporal and spatial locality were followed closely: the Maxeler approach decouples temporal and spatial activities, continues to run temporal activities on a controlflow host, and moves spatial activities to a dataflow accelerator. The fact is that the teaching process is more effective if the subject matter is taught more gradually, starting from the well-known facts, until a clear need for a new paradigm is announced [16]. As it will be shown later, the WebIDE tool starts from classical Eclipse notions and leads the student/user to the most sophisticated aspects of dataflow programming techniques. Similarly, examples of the AppGallery start from the well-known mathematical models of algorithms ported in the dataflow environment; then we show, step by step, how to port various language constructs from the controlflow environment into the dataflow environment.

Finally, the architecting of the system was done in such a way so that sophisticated educational concepts could be addressed using simple implementation steps. This is especially important if the set of tools has to be used in one specific application domain, like teaching developers of medical or financial systems to use dataflow technology.

On the AppGallery website, every application is represented by objects that carry both educational content and concrete education-oriented information (the following five issues are considered crucial for the presentation of an application within the AppGallery system):

- Image (like in an art gallery);
- Name of the application;
- Short summary of the application;
- Authors' names;
- A number of colorful interactive buttons located below the image.

The Chinese proverb "One Picture is Worth Ten Thousand Words" refers to the same guiding principle used for presenting applications in the AppGallery: a complex idea can be conveyed with a single still image. Colorful and interactive buttons located below the application image are designed to invite a visitor to explore and interact with the application, while at the same time providing the visitor with useful information about the application. The main objective is to bring the user as quickly as possible into the essence of the application and its dataflow implementation.

The application is presented with: (1) image; (2) name; (3) short summary; (4) authors; and (5) colorful interactive buttons, as shown in Fig. 1. The interactive buttons can be active or inactive; they are active if the application has a particular feature. For example, the Classification application, on the right, has the DFE SRC button active, while the Correlation

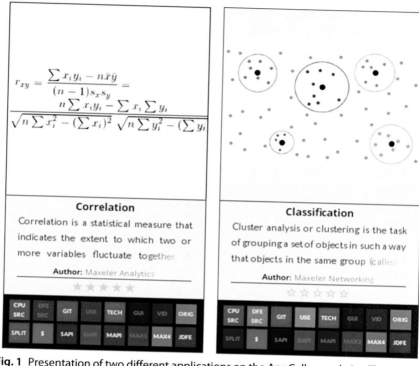

Fig. 1 Presentation of two different applications on the AppGallery website. The 16 buttons refer to the following: (a) controlflow source code; (b) dataflow source code; (c) link to source code on GitHub; (d) application use cases; (e) technical documentation; (f) graphical user interface; (g) video; (h) original, unaccelerated, code; (i) accelerated code; (j) author contact information; (k) SAPI interface; (l) DAPI interface; (m) MAPI interface; (n) support for MAX3 dataflow system; (o) support for MAX4 dataflow system; and (p) support for JDFE dataflow system.

application, on the left, has the DFE SRC button inactive. This means that the authors of the Classification application have provided the source code of the DFE as a part of their application, while the authors of the Correlation application have not provided the DFE source code of their application.

The major buttons are: (1) GitHub, with the code to use; (2) USE, which is a powerpoint with information on the input/output data sets and internal algorithm/method; (3) TECH, which is a powerpoint with implementation-related details. Other buttons are: (4) CPU SRC, controlflow source code; (5) DFE SRC, dataflow source code; (6) GUI, graphical user interface of the application; (7) VID, video of the application; (8) ORIG, original code of the application; (9) SPLIT, division of application into controlflow part and dataflow part; (10) $, commercial information;

(11) SAPI, single action API; (12) DAPI, Dataflow Engine API; (13) MAPI, multiple action API; (14) MAX3, support for MAX3 cards; (15) MAX4, support for MAX4 cards; and (16) JDFE, support for the Maxeler-Juniper switch.

In addition to all the abovementioned, the AppGallery is connected with the following essential parts of the Maxeler dataflow system:

- WebIDE and MaxIDE, which enable modifications of existing designs and/or creations of new designs correlated with existing designs;
- MaxelerOS, which enables integration of the designs made with WebIDE and MaxIDE with the underlying hardware infrastructure, so the speed and comfort of the operation are maximized;
- Educational Tutorials, which teach the users/students how to develop their own applications, by invoking the existing applications via the MaxelerOS, and by enabling the users/students to edit and modify the existing applications using the WebIDE/MaxIDE environment;
- Educational Textbooks, which shed light on the essential notions of dataflow supercomputing, so the users/students could acquire the fundamentals, which enables them both to understand the essence and to create the innovations that would make new advancements in the concept itself.

The integration of all the above is of crucial importance for the mission of the AppGallery, so the design of the AppGallery infrastructure had to take into careful consideration the integration mechanisms, which will be addressed later in this text.

The field of dataflow is rapidly growing, so any supporting infrastructure has to be *easy to expand* with new functionalities in the qualitative way, and *easy to scale* in the quantitative way, since the AppGallery could grow in size, as the dataflow applicability is without limits (the number of algorithms that could benefit from the dataflow concept is unlimited).

To achieve the expandability and the scalability goals, the infrastructure had to be set in a way which is fully modular (in a top–down fashion) and well structured (in a bottom–up fashion). For future expandability, we also had to take into consideration the possible challenges of future computing systems, scientific challenges, and beyond [17–19].

The key issue for the speed of future new developments and for the programmer productivity in future developments probably is the fact that the used technology has to be based on the simplest possible structures that the used technology supports. In other words:

The simpler the mechanism used for the AppGallery implementation, the more successful is the strategy which was set to be in the fundamentals of the architecting and the implementing processes.

The bottom line is that the simpler the notions of the designs described next, the more successful the AppGallery project.

Before the details of the AppGallery project are presented, there is an overview of the general dataflow concept (Section 2), as well as a number of separate overviews, related to the Maxeler Technologies mission in education (Section 3), the Maxeler Technologies approach to dataflow (Section 4), and the methods to accelerate applications using the Maxeler approach (Section 5). Architecting the AppGallery is presented in Section 6, types of the AppGallery users are defined in Section 7, and the AppGallery processes are defined in Section 8. Implementation details are given in Section 9 (although the implementation is fairly sophisticated, this section is written in a way which is easy to comprehend due to the modular architecture used and due to the fact that all crucial issues are defined in earlier sections).

Finally, success measures are given in Section 10 and conclusions are given in Section 11.

2. ABOUT THE DATAFLOW CONCEPT

The dataflow concept is not new. Dataflow hardware architectures were a very popular research topic in the 1970s and early 1980s. Jack Dennis, from MIT, introduced the static dataflow architectures [20,21] and Arvind [22], also from MIT, introduced the dynamic dataflow architectures. Early dataflow machines did not have much commercial success, perhaps due to the fact that they were targeting the general purpose processing arena, and that reconfigurable hardware was not effective.

Recent technology advancements, like modern FPGA (Field-Programmable Gate Arrays) chips with a large number of programmable elements, embedded BRAM, and embedded DSP blocks, enabled successful implementation of dataflow machines for acceleration purposes.

One way to understand the essence of dataflow-based acceleration of controlflow applications is to study the work of the Nobel Laureate Ilya Prigogine, who claims that the entropy of the system could be, under certain conditions, decreased by injecting energy into the system. The work of Milutinovic [23] argues that the split of temporal and spatial computing lowers the entropy of a computing system. Splitting the computation between controlflow and dataflow paradigms results in lower system entropy, and consequently in higher potentials, as regards power, speedup, and complexity.

Another way to understand the essence of dataflow computing is to study the work of the Nobel Laureate Richard Feynman, who claims that, under certain conditions, arithmetic and logic computations could be done with zero energy, while the communication between the processor and memory, or between two processors, takes energy that diverges to infinity, as the length of communication distance keeps increasing. All this paves the way for the success of dataflow based on generation of execution graphs and their mapping onto appropriate hardware structures [24].

One way to understand the essence of the effort to generate a dataflow execution graph with the minimal length of edges is to study the work of Kolmogorov, who defined the so-called Kolmogorov complexity—the minimal complexity needed to move information from point A to point B. The Maxeler Technologies approach to the generation of execution graph follows the notions of the Kolmogorov theory [4].

A way to understand the essence of the effort to utilize the internal execution graph pipelines is to study the work of Cutler of IBM [25]. Cutler claims that latency and precision could be traded. In other words, if internal pipelines are deeper and data more remote in space (this is programming in space), the precision that one could achieve is higher.

3. ABOUT THE MAXELER TECHNOLOGIES MISSION IN EDUCATION

Maxeler Technologies has more than 10 years of experience in producing production-ready dataflow hardware solutions, software solutions, and development tools for programming dataflow machines. Maxeler Technologies core expertise is delivering complete high-performance dataflow solutions and platforms in a wide range of application domains like: Oil & Gas, Finance Analytic, Low-Latency Networking, High-Frequency Trading, Security, and Scientific Computing [4]. Some of the biggest companies that have purchased the Maxeler dataflow systems are Chevron, Schlumberger, J.P. Morgan, and Chicago Mercantile Exchange (CME) [26,27].

Compared to controlflow approaches, the Maxeler dataflow approach has another two important advantages: it is better suited for approximative computing and it enables effective trade-off between latency and precision.

The results of the work of the Nobel laureate Kahneman [28] indicate that there is a large body of applications for which approximative computation is good enough for proper decision making. The resources saved by selecting the approximative computation could be reinvested into improvements of other, more important kind.

The results of the work of Nobel laureate Novoselov [29] indicate that there is need for computing structures that could enable trade-offs between latency and precision. Applications that tolerate higher latencies between input and output could achieve more precision with less hardware resources.

The educational mission of the Maxeler AppGallery project is considerably enhanced by the fact that the selected application examples (for presentation to public in general and students in particular) utilize the wisdom coming from the research of the above referenced four Nobel laureates. By selecting the examples dealing with spatial and temporal locality, with decreased entropy, with approximative computing, and with trade-offs between latency and precision, the educators who use AppGallery are equipped with tools that teach fundamental issues through simple examples: (a) for each application, its execution graph is shown, from which one can study the relationships (for that application) between the length of graph edges and graph topology, which is determined by the position of graph nodes responsible for arithmetic and logic (e.g., Linear regression application); (b) for most applications, it is shown how the temporal and spatial data are decoupled, so that the Maxeler compiler is able to produce the most effective execution graph (e.g., Classification application); (c) many applications deal with approximative computing and demonstrate superiority when floating point numbers are transformed into fixed point numbers (e.g., Reverse Time Migration application); and (d) in some applications, one can introduce longer latencies between input and output, or a higher bandwidth of the path from input to output, which could result in more precision with less hardware resources (e.g., Network Latency Measurement application).

3.1 About MAX-UP

MAX-UP is the University Program of the company Maxeler Technologies. The goal of the MAX-UP program is to provide universities and research institutions with state-of-the-art dataflow software and hardware, enabling education of students on how to write HPC dataflow applications and enabling researchers to use the fastest programmable technology for high-end data-intensive HPC applications in their research. At the time of writing of this chapter, the MAX-UP program had over a hundred universities around the globe as its members. Fig. 2 shows the universities participating in the program on the world map. MAX-UP provides its members with exclusive hardware (like the GALAVA PCIe cards [30]), academic

Fig. 2 The map of locations of the MAX-UP members around the world.

pricing, free access to Maxeler software, and free access to Maxeler educational material.

The Maxeler educational material includes the following tutorials:
- MaxCompiler Tutorial: Multiscale Dataflow Programming;
- MaxCompiler Tutorial: Kernel Numerics;
- MaxCompiler Tutorial: Manager Compiler;
- MaxCompiler Tutorial: Dataflow Programming for Networking;
- MaxCompiler Tutorial: State Machine;
- Acceleration Tutorial: Loops and Pipelining;
- MaxGenFD Tutorial;
- Maxeler DFE Debugging and Optimization Tutorial.

The Maxeler-extended educational material also includes the books published by Springer and Elsevier [31,32], as well as related references [21,33].

4. ABOUT THE MAXELER TECHNOLOGIES APPROACH TO DATAFLOW

Maxeler Technologies produces dataflow accelerators that contain one or more dataflow engines (DFEs) with large amounts of RAM. Some dataflow accelerators also have networking ports attached directly to the DFEs. The latest generation of Maxeler accelerators, at the time of writing of this chapter, is the MAX4 series.

The MAX4 series of accelerators is based on Altera FPGA chips and consists of CORIA, MAIA, ISCA, JDFE, and GALAVA cards. The ISCA and JDFE cards, shown in Fig. 3, have networking capabilities and are often referred to as MAX4N cards. GALAVA, shown in Fig. 4, is an affordable PCIe card available exclusively to the MAX-UP members.

Fig. 3 The MAX4N JDFE Networking card.

Fig. 4 The GALAVA PCIe card. The GALAVA is an affordable dataflow accelerator, available exclusively to the members of the MAX-UP program.

Dataflow programs that can be executed on the DFEs are written in the programming language MaxJ. Dataflow programs written in MaxJ execute in space, in contrast to programs written for CPU which execute in time. The difference is best explained by analogy, comparing a factory production line (DFE) and a skilled worker (CPU). A skilled worker can only do one thing at the time, and always occupies the same amount of space, while all the workers at the factory production line do the work in parallel and the factory production line could be of various lengths, depending on the product that the factory is producing.

In essence, in dataflow computing, programs are written to configure the hardware, not to control the flow of data through the aforementioned execution graph.

The execution graph is generated by the Maxeler Compiler. The output of the Maxeler Compiler is then transformed into the .MAX file, a binary file created with the help of the synthesis tool of the provider of FPGA chips.

4.1 About the MaxJ Language

MaxJ is a high-level programming language, used for programming Maxeler dataflow machines, with syntax similar to the Java syntax, as indicated in Fig. 5. MaxJ is a DSL (domain-specific language) and represents a superset of the classical Java. Extensions of classical Java exist in two dimensions: (a) a set of over 100 built-in Java classes is added to assist the programmer in space, and (b) a new set of variables and operators is added—the DFE variables and operators. Variables of classical Java serve to inform the compiler what to do. The DFE variables physically flow through the hardware. MaxJ is one implementation of the Open SPL

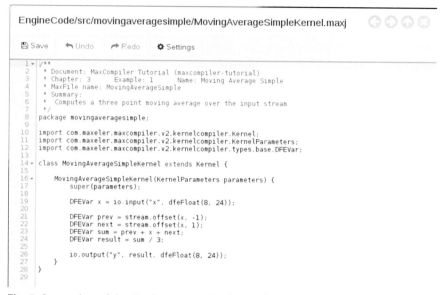

EngineCode/src/movingaveragesimple/MovingAverageSimpleKernel.maxj

```
1   /**
2    * Document: MaxCompiler Tutorial (maxcompiler-tutorial)
3    * Chapter: 3        Example: 1      Name: Moving Average Simple
4    * MaxFile name: MovingAverageSimple
5    * Summary:
6    *   Computes a three point moving average over the input stream
7    */
8   package movingaveragesimple;
9
10  import com.maxeler.maxcompiler.v2.kernelcompiler.Kernel;
11  import com.maxeler.maxcompiler.v2.kernelcompiler.KernelParameters;
12  import com.maxeler.maxcompiler.v2.kernelcompiler.types.base.DFEVar;
13
14  class MovingAverageSimpleKernel extends Kernel {
15
16      MovingAverageSimpleKernel(KernelParameters parameters) {
17          super(parameters);
18
19          DFEVar x = io.input("x", dfeFloat(8, 24));
20
21          DFEVar prev = stream.offset(x, -1);
22          DFEVar next = stream.offset(x, 1);
23          DFEVar sum = prev + x + next;
24          DFEVar result = sum / 3;
25
26          io.output("y", result, dfeFloat(8, 24));
27      }
28  }
29
```

Fig. 5 Screenshot of the *MovingAverageSimpleKernel* written in the MaxJ language. The *MovingAverageSimpleKernel* calculates the moving average of the input stream on the window size of three elements.

(Spatial Programming Language) [34]. MaxJ implements a subset of functionalities defined by the Open SPL specification.

4.2 Accessing DFE

A program executing on the CPU needs to use SLiC (Simple Live CPU) interface to access a DFE. Every DFE can be accessed over Basic Static, Advanced Static, or Dynamic SLiC interfaces.

MaxSkins are wrappers around SLiC C interfaces that allow a DFE to be accessed from different programming languages, like Python and Java. MaxSkins have been developed for almost all popular high-level languages.

The DFE are connected to the controlflow host via the PCIe bus (for smaller systems) or via the InfiniBand bus (for bigger systems).

5. ABOUT THE METHODS FOR ACCELERATING APPLICATIONS USING MAXELER DATAFLOW TECHNOLOGY

Refs. [28,35] show that speedups of 10–100× and power reductions of 10–100× using the same amount of physical space could be achieved when an application is accelerated using Maxeler dataflow technology. When accelerating applications using Maxeler dataflow technology, it is important to understand that not all parts of the application are suitable for dataflow acceleration. The first step of the acceleration process should be splitting the application into two parts: one which controls the flow of data (controlflow) and one which does some processing on the stream of data (dataflow). The first one is usually made up of conditions and control logic, while the second one is usually made up of kernels (loops processing data streams).

The dataflow part of the application is the part suitable for acceleration on the Maxeler dataflow machines. Kernels that are going to be accelerated need to be implemented in the MaxJ language. The code blocks which determine how Kernels are connected to each other, to the memory, and to the CPU, so-called Managers, also need to be implemented in the MaxJ language. Instructions on how to write Kernels and Managers in MaxJ can be found in MaxCompiler tutorials [36].

Maximal acceleration is obtained if the programmer in space is able and willing to invest time and expertise in the following domains:

- Appropriate algorithmic changes;
- Appropriate changes of the input/output data choreography;
- Appropriate utilization of internal pipelines;
- Appropriate utilization of the PCIe bus;
- Appropriate utilization of the required floating-point precision, through the conversion of the calculation from floating point to fixed point, which is an application-specific issue.

Only the synergistic interaction of all the above can ensure the maximal performance of the dataflow approach.

There are two steps in the process (as shown in Fig. 6):

1. In the first step, the original application is split into two parts:
    a. Controlflow part
    b. Dataflow part
2. In the second step, the dataflow part of the split application is moved to the DFE by implementing it in the MaxJ language.

There are three steps in the kernel acceleration process (as shown in Fig. 7):

1. The first step is to write MaxJ implementation of the Kernel and the Manager.
    a. Kernel is an implementation of the accelerated function.
    b. Manager is responsible for internal data communication.
2. The second step is to compile MaxJ files into the .MAX file, using MaxCompiler.

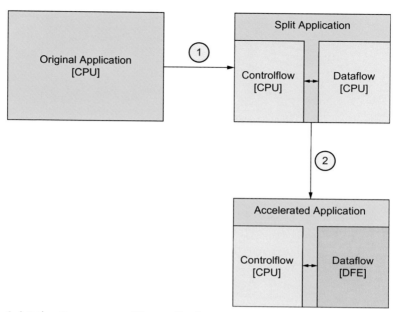

Fig. 6 Acceleration process of the application.

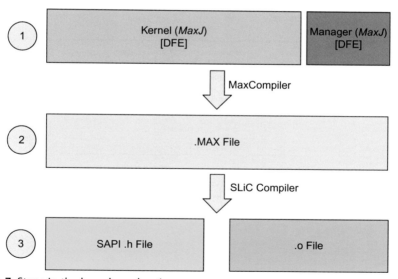

Fig. 7 Steps in the kernel acceleration process.

3. The third step is to compile the .MAX file into .h and .o files that are going to be used by the accelerated application.

Note that the third step is not needed if the accelerated application uses Dynamic interface to access DFEs (more information can be found in the MaxCompiler tutorial [33]).

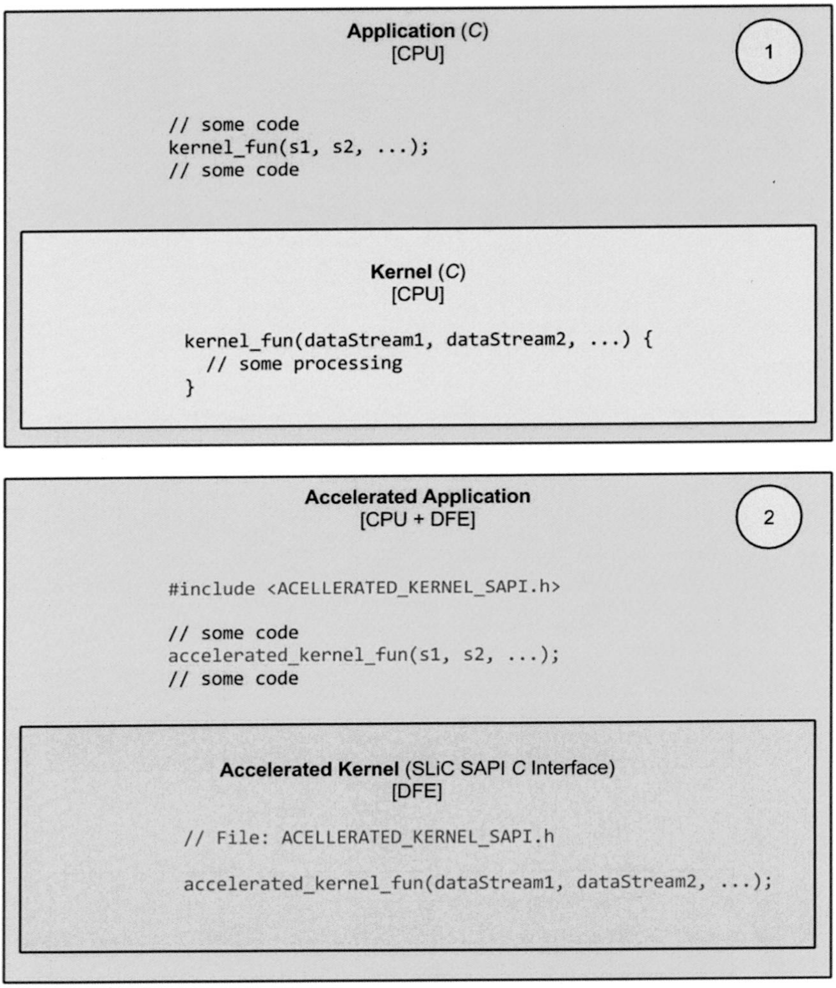

Fig. 8 This figure shows typical code change that needs to be done when using SLiC SAPI (Single Action API) interface to communicate with the DFEs, in order to use the accelerated kernel instead of the original kernel: (1) original application and (2) accelerated application.

The infrastructure presented in Figs. 7–9 was used to support two European projects from the series FP7: (a) ARTreat [37], which was used to data mine correlations between the anamnesis related to lifestyle, genetic material described by 40 DNA parameters, and the standing of carotids and (b) ProSense [38], which was used to data mine the information obtained by wireless sensor networks deployed for the purpose of public health enhancement. In the first case, the purpose was diagnosed on the

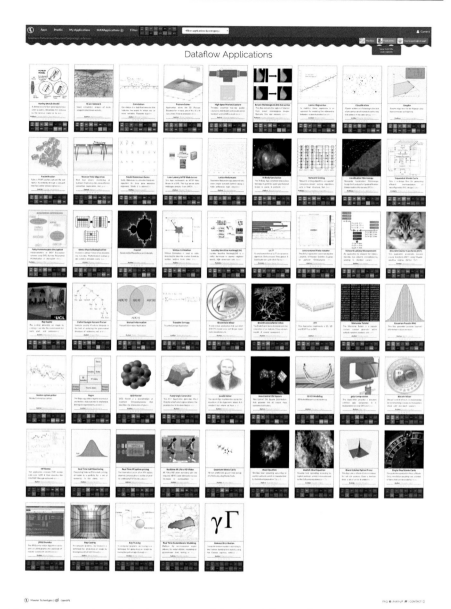

Fig. 9 A screenshot of the homepage of the AppGallery website.

individual level. In the second case, the goal was the detection of threats on the public level. In both cases, sophisticated algorithms had to be used, and control theory was not able to provide solutions for online data, so the dataflow paradigm had to be used.

Fig. 10 Screenshot of the navbar for Non-Registered User.

6. ARCHITECTING THE AppGallery

The AppGallery is a web portal, publicly accessible at http://appgallery. maxeler.com, which allows its visitors to browse the collection of existing dataflow applications and learn more about them. On one hand, the AppGallery is a good starting point for people who are not experienced with dataflow programming, because they can explore what others have done and they can look at the source code and documentation of the many open-sourced applications in the AppGallery. On the other hand, people who have more experience with dataflow can use the AppGallery to keep up to date with what other developers are doing and also rate their applications.

The AppGallery allows application developers to share their work with the rest of the community. Fig. 10 shows how applications are presented using the AppGallery. Authors can choose if they want to open-source their applications, or some of its parts, or not. There is an option for interested users to contact the author of the application with the request for commercial licensing.

The AppGallery examples, as indicated before, could be linked with the WebIDE tool for further developments and for educational purposes—the instructor is able to edit the examples, to analyze the impact of changes, and to discuss alternatives.

7. DEFINING THE TYPES OF THE AppGallery USERS

There are eight different types of users on the AppGallery:
1. Non-Registered User ($NREG$);
2. Registered User (REG);
3. Developer (DEV);
4. Principal Investigator (PI);
5. Principal Investigator Administrator (PI-A);
6. User Administrator (UA);
7. Content Administrator (CA);
8. Super Administrator (SA).
The rest of this section will provide a detailed explanation of each user type.

7.1 Non-Registered User

A Non-Registered User is anybody who stumbles upon the AppGallery website and does not wish to register.

Non-Registered Users can only see the homepage with the list of published applications and with limited details about each application.

Non-Registered Users can see all the published applications in the AppGallery and basic information about the applications, as shown in Figs. 11 and 12. Every app in the AppGallery has (1) an image, (2) a short description, (3) current rating, and (4) app components, as discussed in Section 1.

Clicking on the application image or on the application description will open a pop-up modal window and will show more details about the application.

7.2 Registered User

A Registered User is anybody who registered for the AppGallery website in order to be able to see more information about the applications in the gallery or to download application executables.

Registered Users can log onto the AppGallery website from the login page, as seen in Fig. 13. The login page could be accessed from the homepage by clicking on the *Login* link at the upper right corner of the page.

Registered Users can see the homepage, their profile page (as shown in Fig. 14), full details about all published applications (as shown in Fig. 15), and profile pages of other users (as shown in Fig. 16).

Registered Users have an additional option to rate every app on the AppGallery compared to Non-Registered Users, as shown in Fig. 15.

Registered Users can access their profile page by clicking on the *Profile* button in the navbar. The profile page contains the list of apps that a user has created, as shown in Fig. 16. This page is not public so only logged-in users can see it.

7.3 Developer

A Developer is anybody who wishes to submit their applications to the AppGallery. Developers can do all the things that Registered Users can do, but in addition Developers can create and publish new applications, as shown in Fig. 17. All applications are reviewed before they are publicly visible by the AppGallery content administrators.

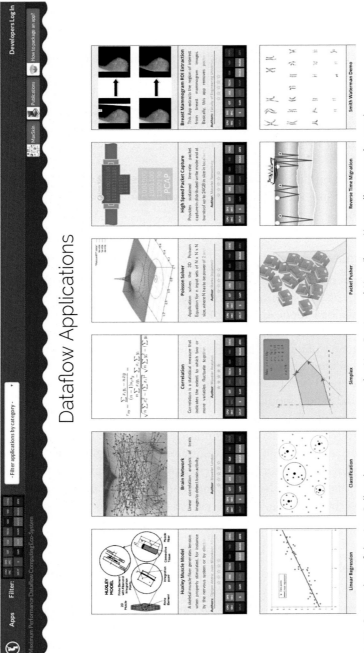

Fig. 11 Screenshot of the AppGallery homepage as seen by Non-Registered User (http://appgallery.maxeler.com).

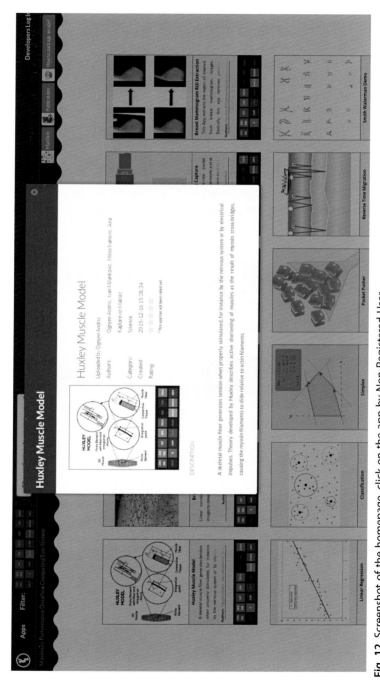

Fig. 12 Screenshot of the homepage, click on the app by Non-Registered User.

Fig. 13 Screenshot of the Login page (http://appgallery.maxeler.com/#/login).

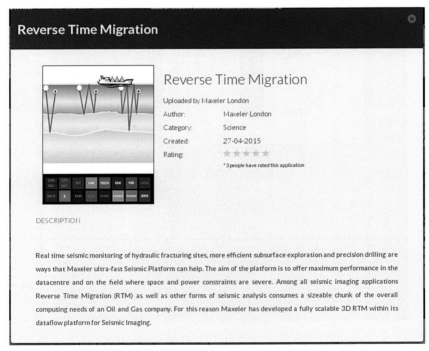

Fig. 14 Screenshot of the navbar for Registered User. The Registered User has a profile link in their navbar that allows the user to see and edit his profile information. In the *upper right corner*, there is the identification of the currently logged-in user.

Reverse Time Migration

Reverse Time Migration

Uploaded by Maxeler London

Author: Maxeler London

Category: Science

Created: 27-04-2015

Rating: ★★★★★

*3 people have rated this application

DESCRIPTION

Real time seismic monitoring of hydraulic fracturing sites, more efficient subsurface exploration and precision drilling are ways that Maxeler ultra-fast Seismic Platform can help. The aim of the platform is to offer maximum performance in the datacentre and on the field where space and power constraints are severe. Among all seismic imaging applications Reverse Time Migration (RTM) as well as other forms of seismic analysis consumes a sizeable chunk of the overall computing needs of an Oil and Gas company. For this reason Maxeler has developed a fully scalable 3D RTM within its dataflow platform for Seismic Imaging.

Fig. 15 Screenshot of the homepage, click on the app by the Registered User.

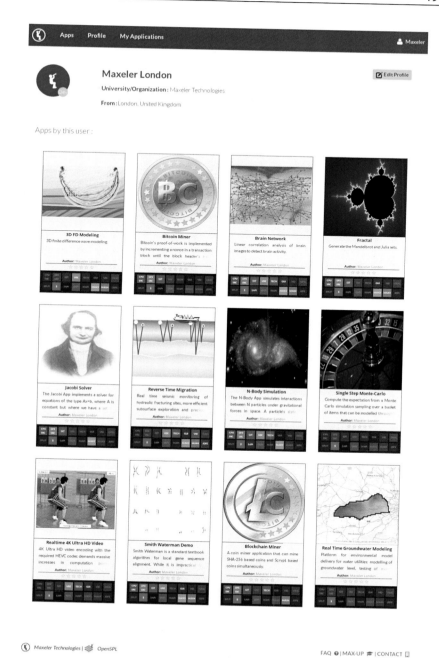

Fig. 16 Screenshot of the Registered User profile page (URL: depends on the user; http://appgallery.maxeler.com/#/user/<user>).

Fig. 17 Screenshot of the navbar for Developer. The Developer has *MyApplications* link in their navbar that link allows them to manage all of their applications and to upload and publish new application to the AppGallery website.

My Applications

You can manage your applications from here.

List of your apps:

3D FD Modeling	✏️ Edit ⊘
Bitcoin Miner	✏️ Edit ⊘
Brain Network	✏️ Edit ⊘
Jacobi Solver	✏️ Edit ⊘
Fractal	✏️ Edit ⊘
Single Step Monte-Carlo	✏️ Edit ⊘
N-Body Simulation	✏️ Edit ⊘
Reverse Time Migration	✏️ Edit ⊘
Fast Fourier Transform 1D	✏️
Fast Fourier Transform 2D	✏️
Smith Waterman Demo	✏️ Edit ⊘
Realtime 4K Ultra HD Video	✏️ Edit ⊘
Blockchain Miner	✏️ Edit ⊘
Real Time Groundwater Modeling	✏️ Edit ⊘

+ New application

Maxeler Technologies | OpenSPL   FAQ ❓ | MAX-UP 🎓 | CONTACT 📇

Fig. 18 Screenshot of the My Apps page (URL: depends on the user; http://appgallery. maxeler.com/#/user/<user>/myapps).

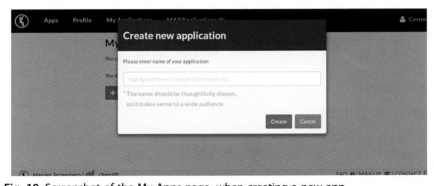

Fig. 19 Screenshot of the My Apps page, when creating a new app.

Fig. 20 Screenshot of the navbar for Principal Investigator and Principal Investigator Administrator. The PI and PI-A have *MyDevelopers* link in their navbars: that link allows them to manage their developers, approve or decline developers that are on their pending lists, and invite new developers.

Fig. 21 Screenshot of the Developer Approval Page (URL: different for each PI; http:// appgallery.maxeler.com/#/user/<PI>/myDevelopers).

By clicking on the *MyApplications* button in the navbar, Developers can (1) access a list of all their applications; (2) monitor the state of their applications; (3) monitor the outcome of the application review; and (4) create and publish new applications, as shown in Figs. 18 and 19.

7.4 Principal Investigator and Principal Investigator Administrator

A Principal Investigator is usually either: (1) a team leader at the company, (2) the main researcher at a research institution, or (3) a professor at the university. Their main role is to manage their staff/students. The Primary Investigator Administrator is usually either (1) a secretary, (2) a research assistant, or (3) a teaching assistant.

Principal Investigators and Principal Investigator Administrators can do all the actions that Registered Users can do, plus they can manage their developers, as shown in Fig. 20.

By clicking on the *MyDevelopers* button in the navbar, PI and PI-A can see the list of their active developers, invite new developers, see the list of pending developers, and approve or decline developers from their pending list of developers, as shown in Fig. 21.

7.5 User Administrator, Content Administrator, and Super Administrator

There is no registration process for UA, CA, and SA types of users. These users are automatically created when the AppGallery website starts

Fig. 22 Screenshot of the navbar for User Admin. The User Admin has *MaxUsers* link in their navbar, which allows them to manage all of the users on the AppGallery website.

Fig. 23 Screenshot of the *MAX Users* page (URL: depends on the user admin; http:// appgallery.maxeler.com/#/userAdmin/<useradmin>/MAXusers).

Fig. 24 Screenshot of the navbar for Content Admin. The Content Admin has *MAX-Applications* link in their navbar, which allows them to manage all of the applications on the AppGallery website.

functioning, so no registration process is needed. The following section gives an overview of the UA, CA, and SA user types.

A User Administrator is responsible for approving and managing PI and Registered User accounts. The UA has full management rights for all accounts on the AppGallery, as shown in Fig. 22.

The User Admin can see four lists: (1) list of active Registered Users, (2) list of pending Registered Users, (3) list of active PIs, and (4) list of pending PIs, as shown in Fig. 23.

A Content Administrator is responsible for all the content (apps) on the AppGallery website. The CA approves every uploaded application. The application approval process is shown in Fig. 24.

By clicking on the *MAXApplications* button in the navbar, the CA can get a list of all pending applications and can Approve or Decline them, as shown in Fig. 25.

A Super Administrator is the only user who has full rights to all the information and the functionality, and is responsible for other administrators (UA, CA), as shown in Fig. 26.

Fig. 25 Screenshot of the *MAXApplications* page.

Fig. 26 Screenshot of the navbar for Super Admin. Super Admin has *MaxUsers* and *Max-Applications* links in their navbar, which allow them to manage all the applications and all of the users on the AppGallery website.

8. DEFINING THE AppGallery PROCESSES

There are seven different processes that can be executed using the AppGallery website:

1. User Registration Process;
2. PI Registration Process;
3. Developer Registration Process;
4. Developer Approval Process;
5. User & PI Approval Process;
6. App Submission Process;
7. App Review Process.

The rest of this section will provide a detailed explanation of each one of the seven different AppGallery processes. For each process, there is a list of actions or steps that need to be performed for successful execution. Every step of process execution will be explained in detail.

8.1 User Registration Process

Anyone can send a request to become a Registered User by filling out a form on the user registration page, as shown in Fig. 27.

Once the registration request is submitted, the UA is notified about the pending registration and can approve or decline the request. When the UA finishes processing the request, the user receives an email with an automatically generated password, as shown in Fig. 28, which can be used by the user to log onto the AppGallery website.

8.2 PI Registration Process

In order to register, the PI needs to send an email to the UA requesting to register as a PI. The UA responds by sending an email with the link to the PI registration page. After this, the PI fills in the PI registration form, as shown in Fig. 29.

The UA is then notified that there is a PI registration request to be reviewed. When the UA approves the request for the PI account creation,

Fig. 27 Screenshot of the User Registration page. Fields marked with *asterisk* (*) are obligatory (http://appgallery.maxeler.com/#/userRegister).

Fig. 28 Screenshot of the received email confirming AppGallery registration approval of the user John Doe. The user is provided with an automatically generated password that he could change during their first log-in.

the PI receives an email with an automatically generated password and a link where developers can register. The PI's Administrator (PI-A) is automatically created and the PI needs to fill in information about the PI-A email address.

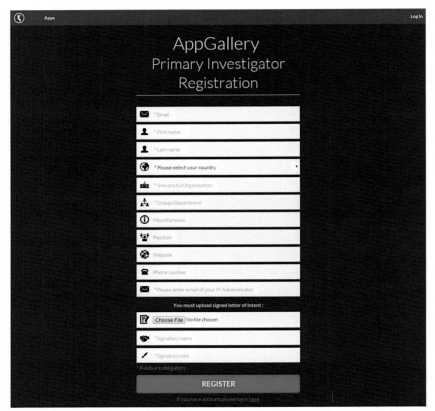

Fig. 29 Screenshot of the PI Registration Page. Fields marked with *asterisk* (*) are obligatory. (URL: The link to the PI registration form is not shown on the AppGallery website, the link to the PI registration form is sent by User Administrator. The link to be sent: http://appgallery.maxeler.com/#/PIregistration.)

8.3 Developer Registration Process

In order for developers to register they first have to receive the registration link from their PI or PI-A. On this link, the developer has to fill in an online developer registration form, as shown in Fig. 30.

Once this has been finished, the PI or PI-A is notified about the registration and can approve the request. When the PI or PI-A finishes processing the request, the developer receives an email with an automatically generated password which can be used to log into the AppGallery.

8.4 Developer Approval Process

The PI or PI-A can see the list of pending developers with all details about a new user and approve or decline their registration request, based on those details.

Fig. 30 Screenshot of the Developer Registration page. Fields marked with *asterisk* (*) are obligatory. (URL: No static link; developer gets the registration link from their PI or PI-A.)

Fig. 31 Screenshot of the Developer Approval Page (URL: different for each PI; http://appgallery.maxeler.com/#/user/<PI>/myDevelopers).

The PI and PI-A have a possibility to approve or decline more than one developer at once, as shown in Fig. 31. To approve or decline more than one developer at a time, select all desired developers for Approval/Decline and click on the Approve or Decline button. A pop-up will show how many developers have been selected for approval, how many for decline, and ask for confirmation. Once confirmed, all the selected developers will be Approved/Declined.

8.5 User & PI Approval Process

User administrator can look through the list of pending users and PIs, as shown in Fig. 32. By clicking on the "View signature" button, next to

Fig. 32 Screenshot of the User & PI Approval page (URL: depends on the user admin; http://appgallery.maxeler.com/#/userAdmin/<useradmin>/MAXusers).

the PIs, the administrator can see: (1) signatory name; (2) signatory role; and (3) download scanned letter of intent for the current user.

The user administrator can select which users are for approval and which are for decline, and when ready can approve or decline them by clicking on the Approve/Decline button. Once the Approve/Decline button has been clicked, a modal pop-up appears with a question to confirm this decision and after clicking OK, users/PIs are approved/declined.

8.6 App Submission Process

In order for an app to be submitted, it needs to have a unique name and a form for uploading application has to be filled out. The form consists of (1) the upload icon, (2) description and additional info, and (3) documentation and archive, as shown in Fig. 33.

Again, issues related to extensibility and scalability are given top priority.

At any moment, the developer can save the current work and finish the application creation at any time by clicking on the *SAVE* button, as shown in Fig. 33.

Once the developer finishes filling in the app submission form, the application can be published, as shown in Fig. 34.

8.7 App Review Process

Content administrator can see a list of pending applications with a short description of every application. By clicking on the application icon, a modal pop-up appears with further application details and an Accept/Decline option, as shown in Figs. 35 and 36.

If the CA decides to decline the application, he/she can write feedback to the developer where he/she could state reasons for declining the application and possible improvements to the application, as shown in Fig. 36. After clicking on the Accept/Decline button in the modal, the modal will close and

Fig. 33 Screenshot of the app Submission page.

Fig. 34 Screenshot of the app Submission page, publish confirmation.

this application will be checked on the list for decline/approval. Finally, the CA can approve/decline all the processed applications by clicking on the Approve/Decline button. After clicking on the Approve/Decline button, the CA would be asked to confirm his/her decision. Once confirmed, all the applications selected for approval will be accepted and the ones selected for decline will be declined.

In case that the application has been declined, the developer receives an email saying that his/her application has been declined and that he/she has to log in to see more information. The developer goes to My Applications page, enters the application that has been declined, and then he/she can see content administrator feedback as shown in the screenshot. If the application is approved, it becomes visible on the homepage and on the author profile page.

9. IMPLEMENTATION DETAILS

This section covers the following issues: system overview, client component, server component, technologies used, and success measures.

9.1 System Overview

This section gives the implementation details, and shows how the initial requests were fulfilled, as far as the synergy of presentation, education, design, and testing, where needed.

WebIDE consists of two main components: (1) the client component and (2) the server component. The client and the server components communicate over HTTP RESTful API, as shown in Fig. 37.

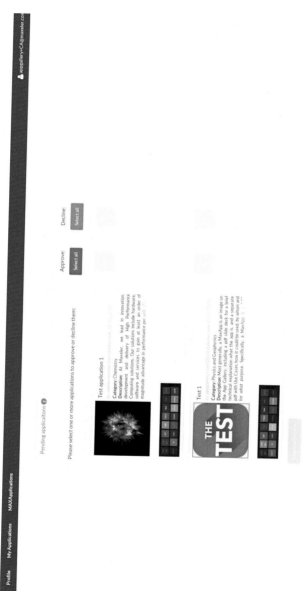

Fig. 35 Screenshot of the MAX Applications page.

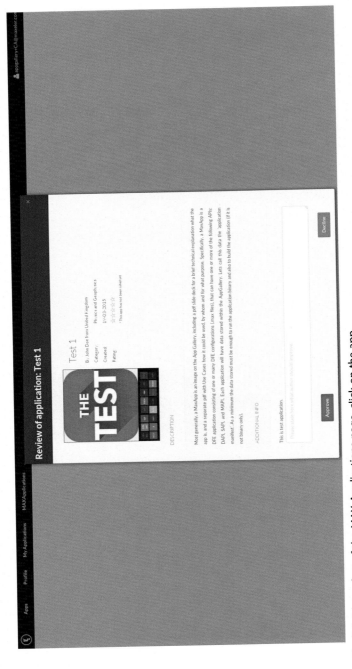

Fig. 36 Screenshot of the MAX Applications page, click on the app.

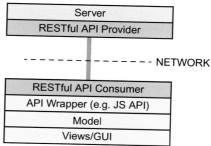

Fig. 37 Overview of the connection between the client and the server component of the system.

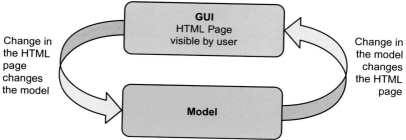

Fig. 38 Explanation of the concept of two-way binding in AngularJS. Every change of data shown on the HTML page (e.g., change of text in a text box) automatically changes the data in the model. Every change of data in the model automatically changes the data shown on the HTML page.

9.2 Client Component

The client component gets executed on the client side, inside of the web browser of the visitors of the AppGallery website, and is implemented in the AngularJS framework. The AngularJS framework is an open-source project started by Google. The framework allows developers to quickly develop the client component of the web application. One of the main concepts of the AngularJS framework is two-way data binding. This concept is explained in Fig. 38. The client component consists only of static files (html, javascript, css, and other website assets) that can be served by any web server.

9.3 Server Component

The server component provides RESTful API that the client component uses, and is implemented in Python, using the Flask framework and a couple of framework extensions (see Table 1 for the list of extensions used).

Table 1 Table of Frameworks and Open-Source Projects That Were Used for
Developing the AppGallery

Framework/ Project Name	Short Description	Programming Language
Flask	Microframework for web application development	Python
Flask-RESTful	Flask extension that adds support for building RESTful APIs	Python
Flask-HTTPAuth	Flask extension that adds support for HTTP authentication	Python
virtualenv	Tool for creating isolated Python environments	Python
AngularJS	Framework for developing client side of web applications	JavaScript
Bootstrap	Framework for developing user interface of web applications	LESS, JavaScript
LESS	CSS preprocessor; LESS extends the CSS syntax	JavaScript
AngularUI—Bootstrap	Extension for the AngularJS framework that makes working with Bootstrap easier	JavaScript
pdf.js	PDF reader that works in the Internet browser	JavaScript

Flask is an open-source project started in 2010 by Armin Ronacher. Flask framework allows developers to quickly develop the server component of a web application. The structure of the server component is explained in Fig. 39.

9.4 Technologies Used

Many different open-source projects were used for the development of the AppGallery. Table 1 contains the list of used projects with a short description of each project and the programming language in which the project is implemented.

10. SUCCESS MEASURES

Two experiments were conducted to determine the effectiveness of the tools used. The WebIDE system connected with AppGallery was compared with the MaxIDE system without the AppGallery. The MaxIDE

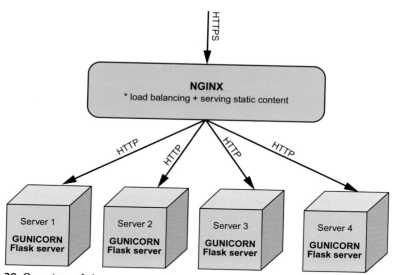

Fig. 39 Overview of the server configuration. NGINX is serving the static files of the AppGallery website and acting as a load balancer. There are multiple gunicorn server instances that are executing the WebIDE server written in the Flask framework. These instances are processing the request they get from the load balancer.

Table 2 Ratios of Time to Design, Program, Test, and Build

	MaxIDE	WebIDE	Ratio
Time to design	$1.20 \times T_{\text{ave}}$	$0.83 \times T_{\text{ave}}$	1.446
Time to program	$1.47 \times T_{\text{ave}}$	$0.68 \times T_{\text{ave}}$	2.162
Time to test	$1.18 \times T_{\text{ave}}$	$0.85 \times T_{\text{ave}}$	1.388
Time to build	$1.02 \times T_{\text{ave}}$	$0.98 \times T_{\text{ave}}$	1.041

The formula means that the results were averaged over a number of individual developers (N). T_i is the time used by an individual developer to perform the specific task.

system is a superset of the WebIDE system regarding the functions it supports; however, it is not connected with the AppGallery and it is not web based.

The following four parameters are relevant for all development efforts: (a) time to design, (b) time to program, (c) time to test, and (d) time to build. The best way to measure the effectiveness of the advanced tool based on synergy compared to the traditional tool based on symbiosis is to provide the ratio of the times the user has spent in one and the other environment. The results are given in Table 2.

Table 3 The Speedup Ratio, Compared to a Control Flow Machine, When WebIDE Was Used (R_{web}) and When the MaxIDE Was Used (R_{max})

	Speedup
R_{max}	2.78
R_{web}	4.62
Ratio	1.66

The parameter R is equal to the ratio of R_{web} and R_{max}.

The central purpose of moving applications from controlflow to dataflow is to gain speedup in conditions where power dissipation is lowered. Therefore, the major question was what speedups the tested students were able to achieve when using WebIDE instead of MaxIDE.

The testing was performed with 20 students, each one working on a different problem (porting a different algorithm from controlflow to dataflow). The results are shown in Table 3 and they are very indicative.

The conclusion from these two experiments is that not only was the time to develop shortened, but also was the time to execute. The first one is important for the project investor, while the second one is important for the project user.

These performance measurements prove that the tools described here were able to enhance the education process at the University of Belgrade, School of Electrical Engineering, Department of Computer Engineering. The same tools were used effectively at Harvard [39] and MIT [40].

11. CONCLUSION

The AppGallery project is still in its early stages and is constantly evolving and improving. With over a hundred universities being members of the MAX-UP program, it is going to be very interesting to see which applications the dataflow developers are going to develop and share with the rest of the community.

The goal of the AppGallery project was, on one hand, to build the best place for dataflow developers to show their work and engage with the rest of community, while on the other hand educating the broader community about various applications that could be implemented using dataflow technology.

However, the prime achievement of this project is the efficient synergy of education, tools, the operating system, and application development. This was achieved through the strategy referred to as: "The simpler the better, or less means more." One of the important benefits of the presented system is its openness to growth. Scientists worldwide are invited to contribute with their own examples of interest for implementation of sophisticated problems that they are working on. We expect the AppGallery ecosystem to grow over time, which enables a specific user community to be created.

The major impact of this work is related to helping the community accept a new paradigm related to programming in space and improved utilization of dataflow-based acceleration technology [41].

REFERENCES

[1] J. Gonzalez, R. Xin, A. Dave, D. Crankshaw, M. Franklin, I. Stoica, GraphX: graph processing in a distributed dataflow framework, OSDI, 2014.
[2] T. Akidau, R. Bradshaw, C. Chambers, S. Chernyak, R. Fern-Moctezuma, R. Lax, et al., The dataflow model: a practical approach to balancing correctness, latency, and cost in massive-scale, unbounded, out-of-order data processing, Proc. VLDB Endowment 8 (12) (2015) 1782–1803.
[3] A. Veen, Dataflow machine architecture, ACM Comput. Surv. 18 (4) (1986) 365–396.
[4] Maxeler, https://www.maxeler.com. accessed on August 1, 2015.
[5] M. Markovic, I. Kostic Kovacevic, B. Nikolic, INSOS—educational system for teaching intelligent systems, Comput. Appl. Eng. Educ. 23 (2) (2015) 268–276.
[6] Z. Stanisavljevic, B. Nikolic, I. Tartalja, V. Milutinovic, A classification of eLearning tools based on the applied multimedia, Multimed. Tools Appl. 74 (11) (2015) 3843–3880.
[7] M. Zivovic, B. Nikolic, R. Popovic, eWISENS: educational wireless sensor network simulator, Int. J. Eng. Educ. 30 (2) (2014) 483–494.
[8] V. Stanisavljevic, V. Pavlovic, B. Nikolic, J. Djordjevic, SDLDS system for digital logic design and simulation, IEEE Trans. Educ. 56 (2) (2013) 235–245.
[9] N. Grbanovic, B. Nikolic, J. Djordjevic, The VSDS environment based laboratory in computer architecture and organisation, Comput. Appl. Eng. Educ. 19 (4) (2011) 685–696.
[10] B. Nikolic, Z. Radivojevic, J. Djordjevic, V. Milutinovic, A survey and evaluation of simulators suitable for teaching courses in computer architecture and organization, IEEE Trans. Educ. 52 (4) (2009) 449–459.
[11] J. Djordjevic, B. Nikolic, T. Borozan, A. Milenkovic, CAL2: computer aided learning in computer architecture laboratory, Comput. Appl. Eng. Educ. 16 (3) (2008) 172–188.
[12] A. Stojkovic, J. Djordjevic, B. Nikolic, WASP—a web based simulator for an educational pipelined processor, Int. J. Electr. Eng. Educ. 44 (3) (2007) 197–216.

[13] J. Djordjevic, B. Nikolic, M. Mitrovic, A memory system for education, Comput. J. 48 (6) (2005) 630–641.

[14] J. Djordjevic, B. Nikolic, A. Milenkovic, Flexible web-based educational system for teaching computer architecture and organization, IEEE Trans. Educ. 48 (2) (2005) 264–274.

[15] B. Davis, D. Sumara, Complexity and Education—Inquiries Into Learning, Teaching, and Research, Routledge, Abingdon, United Kingdom, 2006.

[16] V. Blagojevic, et al., A systematic approach to generation of new ideas for PhD research in computing, Adv. Comput. 104 (2016) 1–19.

[17] Y. Patt, Future microprocessors: what must we do differently if we are to effectively utilize multi-core and many-core chips? IPSI BgD Trans. Internet Res. 5 (1) (2009) 5–10.

[18] M. Perl, Creativity in science, IPSI BgD Trans. Adv. Res. 5 (1) (2009) 2–5.

[19] Friedman, J., "The new views of the universe," IPSI BgD Trans. Adv. Res., Vol. 4, No. 2, pp. 5–7, 2008.

[20] J.B. Dennis, Data flow supercomputers, Comput. Ser. 13 (1980) 48–56.

[21] J. Dennis, Dataflow computation, in: V. Milutinovic (Ed.), Computer Architecture, North Holland, New York, USA, 1988. Chapter 9.

[22] Arvind, M., "Passing the token," Keynote ISCA 2005, 2005.

[23] Milutinovic, V., et. al. "Splitting temporal and spatial computing: enabling a combinational dataflow in hardware," The ACM ISCA Tutorial on Advances in supercomputing, 1995.

[24] M. Flynn, O. Mencer, V. Milutinovic, G. Rakocevic, P. Stenstrom, R. Trobec, et al., Moving from petaflops to petadata, Commun. ACM 56 (5) (2013) 39–43.

[25] V. Milutinovic, Trading latency and performance: a new algorithm for adaptive equalization, IEEE Trans. Commun. 33 (6) (1985) 522–526.

[26] Innovation in Investment Banking Technology, Field Programmable Gate Arrays, J.P. Morgan, New York, 2012.

[27] Dimond, R., Flynn, M., Mencer, O., Pell, O., "MAXware: acceleration in HPC," IEEE HOT CHIPS 20, Stanford, USA, 2008.

[28] D. Kahneman, A. Tversky, Prospect theory: an analysis of decision under risk, Econometrica 47 (2) (1979) 263–292.

[29] N. Kostya, A. Geim, S. Morozov, D. Jiang, Y. Zhang, S. Dubonos, et al., Electric field effect in atomically thin carbon films, Science 306 (5696) (2004) 666–669.

[30] GALAVA—[MAX-UP], https://www.maxeler.com/solutions/universities/galava/. accessed on August 3, 2015.

[31] N. Trifunovic, V. Milutinovic, J. Salom, A. Kos, Paradigm shift in big data supercomputing: dataflow vs controlflow, J. Big Data 2 (2015) 4. May 2015.

[32] A. Hurson, V. Milutinovic (Eds.), DataFlow Processing, Elsevier, Amsterdam, Netherlands, 2015.

[33] Z. Jovanovic, V. Milutinovic, FPGA accelerator for floating-point matrix multiplication, IET Comput. Digit. Tech. 6 (2012) 249–256.

[34] Open SPL, http://www.openspl.org. accessed on February 6, 2016.

[35] V. Milutinovic, J. Salom, N. Trifunovic, R. Giorgi, Guide to DataFlow supercomputing, Springer International Publishing, Switzerland, 2015.

[36] MultiScale DataFlow Programming, Maxeler, Palo Alto, CA, 2015.

[37] The FP7 ARTreat Project, http://www.artreat.kg.ac.rs, 2014.

[38] The FP7 ProSense Project, www.proasense.eu, 2014.

[39] V. Milutinovic, Maxeler Data Flow supercomputing. Invited lectures at Harvard, May 2016.

[40] V. Milutinovic, Maxeler Data Flow supercomputing. Invited lectures at MIT, May 2016.

[41] N. Trifunovic, V. Milutinovic, N. Korolija, G. Gaydadjiev, The Maxeler AppGallery mission in education and research, J. Big Data 3 (1) (2016) 4–34.

ABOUT THE AUTHORS

Nemanja Trifunovic is third year PhD student at the School of Electrical Engineering, University of Belgrade.

During his studies Nemanja did an internship at major tech companies (Google, Microsoft, Maxeler Technologies) every summer and was involved with preparing high school students for IOI (International Olympiad in Informatics), which he participated in himself and had won medals for Serbia during his high school days.

Currently Nemanja is leading a small team of around 15 people in Belgrade working on innovative computing paradigm, DataFlow commuting, as a head of the Belgrade office of the UK company Maxeler Technologies.

Nemanja has participated in a number of conferences and gave a couple of invited lectures. After finishing his bachelor studies Nemanja was awarded "Crown of Success" award for the best student of the Belgrade University.

Boris Perovic received the MSc degree in Computer Science, Internet Computing Specialization from École Polytechnique Fédérale de Lausanne (EPFL), Switzerland. Boris did his master thesis research in Pervasive Parallelism Laboratory, under the supervision of Prof. Kunle Olukotun and Prof. Martin Odersky.

Boris is currently vising research student at Stanford University, USA.

Throughout his studies Boris worked on numerous academic projects and performed several internships. Interning at Net Link Solutions gave him an insight into the fast-moving domain of online commerce and entrepreneurship, but also got him engaged in the open-source world and communities around it. Working at CERN got him in touch with the boundaries of human physics knowledge and scientific computing, and allowed Boris to work with some of the best people in the field.

Petar Trifunovic is student at the University of Belgrade and is working for Maxeler Technologies, UK. He is the author of several DataFlow tools and applications.

Zoran Babovic received the MSc degree in electrical engineering, University of Belgrade, The School of Electrical Engineering, Serbia, in 2004, where he is currently pursuing the PhD degree. He has been working on several research and software development projects, in cooperation with leading EU Institutes and USA/UK companies, such as IPSI Fraunhofer Institute, Germany, Storage Tek, USA, Dow Jones, USA, Maxeler, UK, in the domain of multimedia, data management, dataflow, and real-time software systems. Since 2006, he has been a Research Associate with the Innovation Center, University of Belgrade, The School of Electrical Engineering, Serbia. He participated in four EU FP6 and FP7 research projects in the domain of sensor networks, data mining, and computer architecture, as well as in four innovation and research projects funded by the Serbian Ministry of Education, Science and Technological Development. He has authored several conference and journal papers, and gave numerous talks at conferences in Europe.

 **Ali R. Hurson** is a professor of Departments of Computer Science, and Electrical and Computer Engineering at Missouri S&T. For the period of 2008–2012 he served as the computer science department chair. Before joining Missouri S&T, he was a professor of Computer Science and Engineering department at The Pennsylvania State University. His research for the past 35 years has been supported by NSF, DARPA, Department of Education, Department of Transportation, Air Force, Office of Naval Research, NCR Corp., General Electric, IBM, Lockheed Martin, Penn State University, and Missouri S & T. He has published over 330 technical papers in areas including computer architecture/organization, cache memory, parallel and distributed processing, Sensor and Ad Hoc Networks, dataflow architectures, VLSI algorithms, security, Mobile and pervasive computing, database systems, multidatabases, global information sharing processing, application of mobile agent technology, and object oriented databases.

Professor Hurson has been active in various IEEE/ACM Conferences and has given tutorials on global information sharing, database management systems, supercomputer technology, data/knowledge-based systems, dataflow processing, scheduling and load balancing, parallel computing, pervasive computing, green computing, and sustainability. He served as an IEEE editor, IEEE distinguished speaker, and an ACM distinguish lecturer. Currently, he is Editor-in-Chief of Advances in Computers, editor of The CSI Journal of Computer Science and Engineering, and editor of Computing Journal.

Printed in the United States
By Bookmasters